D0424909

495

Prentice-Hall
Foundations of
Modern Organic Chemistry
Series

KENNETH L. RINEHART, JR., Editor

Volumes published or in preparation

ORGANIC
SYNTHESIS

Robert E. Ireland

Department of Chemistry
California Institute of Technology

PRENTICE-HALL, INC., ENGLEWOOD CLIFFS, N.J.

C—13-640839-7
P—13-640847-8

Library of Congress Catalog Card Number 73-76870
Printed in the United States of America

PRENTICE-HALL INTERNATIONAL, INC., London
PRENTICE-HALL OF AUSTRALIA, PTY. LTD., Sydney
PRENTICE-HALL OF CANADA, LTD., Toronto
PRENTICE-HALL OF INDIA PRIVATE LTD., New Delhi
PRENTICE-HALL OF JAPAN, INC., Tokyo

Current Printing (last digit):
10   9   8   7   6   5   4   3

To an inspiring teacher, William S. Johnson,

and the memory of my Dad

This book was prepared during the tenure of a Fellowship of the Michigan Foundation for Advanced Research, Midland, Michigan, and the author gratefully acknowledges the support of this organization. Without the courageous support and unflagging solace of my wife, Susan, this project would never have reached this conclusion. For this the author is eternally grateful.

Foreword

Organic chemistry today is a rapidly changing subject whose almost frenetic activity is attested by the countless research papers appearing in established and new journals and by the proliferation of monographs and reviews on all aspects of the field. This expansion of knowledge poses pedagogical problems; it is difficult for a single organic chemist to be cognizant of developments over the whole field and probably no one or pair of chemists can honestly claim expertise or even competence in all the important areas of the subject.

Yet the same rapid expansion of knowledge—in theoretical organic chemistry, in stereochemistry, in reaction mechanisms, in complex organic structures, in the application of physical methods—provides a remarkable opportunity for the teacher of organic chemistry to present the subject as it really is, an active field of research in which new answers are currently being sought and found.

To take advantage of recent developments in organic chemistry and to provide an authoritative treatment of the subject at an undergraduate level, the *Foundations of Modern Organic Chemistry Series* has been established. The series consists of a number of short, authoritative books, each written at an elementary level but in depth by an organic chemistry teacher active in research and familiar with the subject of the volume. Most of the authors have published research papers in the fields on which they are writing. The books will present the topics according to current knowledge of the field, and individual volumes will be revised as often as necessary to take account of subsequent developments.

The basic organization of the series is according to reaction type, rather than along the more classical lines of compound class. The first ten volumes in the series constitute a core of the material covered in nearly every one-year organic chemistry course. Of these ten, the first three are a general introduction to organic chemistry and provide a background for the next six, which deal with specific types of reactions and may be covered in any order. Each of the reaction types is presented from an elementary viewpoint, but in a depth not possible in conventional textbooks. The teacher can decide how much of a volume to cover. The tenth examines the problem of organic synthesis, employing and tying together the reactions previously studied.

The remaining volumes provide for the enormous flexibility of the series. These cover topics which are important to students of organic chemistry and are sometimes treated in the first organic course, sometimes in an intermediate course. Some teachers will wish to cover a number of these books in the one-year course; others will wish to assign some of them as outside reading; a complete intermediate organic course could be based on the eight "topics" texts taken together.

The series approach to undergraduate organic chemistry offers then the considerable advantage of an authoritative treatment by teachers active in research, of frequent revision of the most active areas, of a treatment in depth of the most fundamental material, and of nearly complete flexibility in choice of topics to be covered. Individually the volumes of the Foundations of Modern Organic Chemistry provide introductions in depth to basic areas of organic chemistry; together they comprise a contemporary survey of organic chemistry at an undergraduate level.

KENNETH L. RINEHART, JR.

University of Illinois

Contents

PART II—SPECIFIC EXAMPLES

4

WHEREIN THE CARBON SKELETON IS THE THING   55

5

STEREOCHEMISTRY RAISES ITS UGLY HEAD   100

MULTISTAGE SYNTHESIS: LOGISTICS AND STEREOCHEMISTY COMBINE TO PRODUCE NIGHTMARES   123

PART I
The Synthetic Process

The field of organic chemistry is intimately involved with the synthesis of organic molecules and, thereby, the production of new systems for study. The unique property of carbon that allows the formation of carbon-to-carbon bonds opens the door to a vast plethora of theoretically available organic molecules. As the field expands, the compounds that interest organic chemists become more varied. Such compounds run the gamut of man's fancy from naturally occurring molecules to bizarre arrangements of atoms designed to test the ever-evolving tenets of structural theory. Thus, at the heart of organic chemistry is the ability to construct more complex organic molecules from their less complex components by a series of rational procedures. Such is the discipline of synthetic organic chemistry.

Over the years, a large reservoir of organic reactions has been developed; these provide the tools for synthesis. The scope and limitations of these tools are constantly under scrutiny, and many have been honed to superb keenness. However, they remain only the tools of the trade, and a synthetic project can come into being only when a rational stepwise organization of these tools is set down. Thus, the concept of organic synthesis involves the process of arriving at a suitable plan for the interlacing of a series of organic reactions so as to build one molecule from another.

It must be emphasized that a synthetic program of any magnitude is a total organic chemistry experience, and it involves the application of the knowledge and techniques of the entire science. A most thorough knowledge of organic reactions is *de rigueur*. Application of modern spectroscopy is necessary in ascertaining structural changes; knowledge of reaction mechanisms is imperative in interpreting structure-reactivity relationships of proposed stages; mastery of the tenets of structural theory makes possible the evaluation of results in individual changes. Imperative, of course, is familiarity with the laboratory techniques of organic chemistry. Every facet of the realm of organic chemistry will be used by the practitioner of synthetic organic chemistry. It is beyond the scope of this discussion to develop at length *all* of the disciplines that bear on an organic synthesis. The purpose of the discussion is to organize a prior and ever-enlarging knowledge of organic chemistry along the lines of synthesis.

Where do we begin to develop the technique for designing a synthetic program? As any artisan, we must first know our tools well. We must know their capabilities, their generalities, and, especially, their limitations. The ever-changing development of scientific knowledge means this task is never done. New tools are introduced and old ones refined as we become more intimately familiar with the characteristic behavior of the elements. We can, however, divide our tools—old, new, and as yet unknown—into two major classes which will guide our discussion of their use. One group comprises those reactions that allow us to form a new carbon-to-carbon bond, and the other is the set of reactions that provide for changing one functional group into another. The first set of reactions is particularly important because, through the use of these reactions, we can construct our desired carbon framework. We need the second group of reactions to provide sites for new carbon attachments or to prepare a given functional array. When we are familiar with the tools available and their uses, we can pass on to the development of the technique of synthetic planning.

1

Organic Reactions

Probably the aesthetically most satisfying classification of the reactions of organic chemistry is based on reaction mechanism. From this viewpoint, many of the classically famous "name reactions" coalesce into minor variations of a particular reaction type, all of which have the same mechanistic features. A further refinement of these reaction types is useful for a discussion of synthetic organic chemistry. This refinement is a result of the two major problems presented by a synthetic project.

The synthesis of most organic molecules can be broken down into the problem of preparing the carbon framework and that of the introduction, modification, and/or removal of various functional groups. Thus, within the framework of a mechanistic classification, the process of planning a synthetic scheme necessitates a further subdivision of organic reactions into those that are used to form carbon-to-carbon bonds and those that effect functional group changes only. Both of these subclassifications are equally important, since any synthesis will use both types of reactions. From the standpoint of ordering the thought process by which a synthetic scheme arises, this organization emphasizes the different uses to which the reactions are put.

A brief review of the principal carbon-to-carbon bond-forming reactions seems appropriate at the outset. This presentation is not meant to be complete; its purpose is only to suggest a way of organizing information. The following table represents a point of departure from which a more

CARBON-TO-CARBON BOND-FORMING REACTIONS

Type	Representative name reactions
A. 1,2-Carbonyl addition and/or replacement reactions	Grignard reaction, Wittig reaction, aldol-type condensation, cyanohydrin addition, carbene addition
B. Conjugate addition reactions	Michael reaction cyanide addition Grignard addition
C. Olefin addition reactions	Claisen-Cope rearrangement, cycloaddition reactions, Friedel–Crafts reaction, acid catalyzed addition, carbene reactions
D. Displacement reactions	Alkylation (S_N2) reactions, unimolecular (S_N1) alkylation reactions, molecular rearrangements

comprehensive personal compilation may be made. It also serves the purpose of review and codification for further discussion.

The importance of this classification lies in the recognition of the relatively few carbon-to-carbon bond-forming reaction types. Also evident from this list is the central role in synthetic organic chemistry played by the carbonyl group, for the majority of reactions mentioned interact with a carbonyl group at some point.[1]

Consider first the variety of structures possible as a result of the 1,2-addition to a carbonyl group. Both acid (1) and base (2) catalyzed reac-

$$H^\oplus + \;\;\diagdown\!C{=}O \;\; \rightleftharpoons \;\; \overset{\oplus}{\diagdown}\!C{-}OH \;\; \xrightarrow{X^\ominus} \;\; \overset{\overset{\displaystyle X}{|}}{\diagdown}\!C{-}OH \tag{1}$$

$$B^\ominus + \;\;\diagdown\!C{=}O \;\; \longrightarrow \;\; \overset{\overset{\displaystyle B}{|}}{\diagdown}\!C{-}O^\ominus + BH \;\; \rightleftharpoons \;\; \overset{\overset{\displaystyle B}{|}}{\diagdown}\!C{-}OH + B^\ominus \tag{2}$$

tions are available and provide numerous methods for forming new carbon structures. For instance, the carbon skeleton of 3-methylhexane (**3**) is

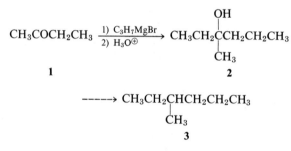

$$CH_3COCH_2CH_3 \xrightarrow[\text{2) } H_3O^\oplus]{\text{1) } C_3H_7MgBr} CH_3CH_2\overset{\overset{\displaystyle OH}{|}}{\underset{\underset{\displaystyle CH_3}{|}}{C}}CH_2CH_2CH_3$$

$$\mathbf{1} \qquad\qquad\qquad\qquad \mathbf{2}$$

$$\text{-----}\rightarrow CH_3CH_2\overset{}{\underset{\underset{\displaystyle CH_3}{|}}{C}}HCH_2CH_2CH_3$$

$$\mathbf{3}$$

available through 3-methyl-3-hexanol (**2**), which in turn arises from the addition of *n*-propylmagnesium bromide to 2-butanone (**1**).

EXERCISE 1

Show how the carbon skeleton of 3-methylhexane may be constructed from 2-butanone by utilization of *two other* type A reactions.

Addition of this Grignard reagent to different carbonyl-containing substrates results in numerous skeletal variations.

[1] The central part played by the carbonyl group in organic chemistry is well documented in C. D. Gutsche, *The Chemistry of Carbonyl Compounds,* in the Foundations of Modern Organic Chemistry Series, Prentice-Hall, Inc., Englewood Cliffs, N.J., 1967.

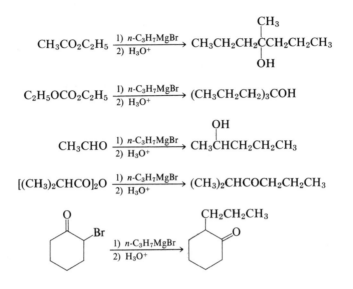

$$CH_3CO_2C_2H_5 \xrightarrow[\text{2) } H_3O^+]{\text{1) } n\text{-}C_3H_7MgBr} CH_3CH_2CH_2\overset{\overset{\displaystyle CH_3}{|}}{\underset{\underset{\displaystyle OH}{|}}{C}}CH_2CH_2CH_3$$

$$C_2H_5OCO_2C_2H_5 \xrightarrow[\text{2) } H_3O^+]{\text{1) } n\text{-}C_3H_7MgBr} (CH_3CH_2CH_2)_3COH$$

$$CH_3CHO \xrightarrow[\text{2) } H_3O^+]{\text{1) } n\text{-}C_3H_7MgBr} CH_3\overset{\overset{\displaystyle OH}{|}}{C}HCH_2CH_2CH_3$$

$$[(CH_3)_2CHCO]_2O \xrightarrow[\text{2) } H_3O^+]{\text{1) } n\text{-}C_3H_7MgBr} (CH_3)_2CHCOCH_2CH_2CH_3$$

EXERCISE 2

Suggest a mechanism for the reaction between 2-bromocyclohexanone and *n*-propylmagnesium bromide that justifies classification under 1,2-carbonyl addition types.

Notice the wide variation of carbon skeletal structures that are available from the 1,2-addition of a nucleophilic species [*n*-C$_3$H$_7$MgBr] to a carbonyl derivative. The important result, for the moment, is not so much the exact product obtained nor the functional system that is generated, but rather the availability of a larger and different carbon skeleton from a smaller, less complex starting material. Thus, through the Grignard reaction with carbonyl derivatives, a synthetic tool is available for the construction of organic molecules.

To further emphasize the utility of 1,2-addition to the carbonyl group as a source of different carbon skeletons, notice several of the other ways that the 3-methylhexane skeleton (**3**) may be prepared.

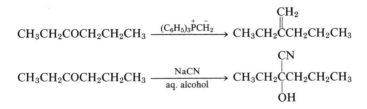

$$CH_3CH_2COCH_2CH_2CH_3 \xrightarrow{(C_6H_5)_3\overset{+}{P}\overset{-}{CH_2}} CH_3CH_2\overset{\overset{\displaystyle CH_2}{\parallel}}{C}CH_2CH_2CH_3$$

$$CH_3CH_2COCH_2CH_2CH_3 \xrightarrow[\text{aq. alcohol}]{NaCN} CH_3CH_2\overset{\overset{\displaystyle CN}{|}}{\underset{\underset{\displaystyle OH}{|}}{C}}CH_2CH_2CH_3$$

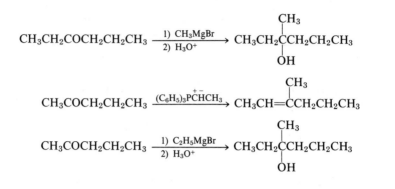

$$CH_3CH_2COCH_2CH_2CH_3 \xrightarrow[\text{2) } H_3O^+]{\text{1) } CH_3MgBr} \underset{\underset{OH}{|}}{CH_3CH_2\overset{\overset{CH_3}{|}}{C}CH_2CH_2CH_3}$$

$$CH_3COCH_2CH_2CH_3 \xrightarrow{(C_6H_5)_3\overset{+}{P}\overset{-}{C}HCH_3} CH_3CH{=}\overset{\overset{CH_3}{|}}{C}CH_2CH_2CH_3$$

$$CH_3COCH_2CH_2CH_3 \xrightarrow[\text{2) } H_3O^+]{\text{1) } C_2H_5MgBr} \underset{\underset{OH}{|}}{CH_3CH_2\overset{\overset{CH_3}{|}}{C}CH_2CH_2CH_3}$$

EXERCISE 3

Show how each of the foregoing products may be converted to 3-methylhexane itself.

In the examples just cited, the carbonyl component varies and, to that extent, dictates the choice of reagent; but in every case, the 3-methylhexane skeleton can be prepared by a 1,2-addition to the carbonyl group. The exact choice of the particular pathway desired depends on several other factors, such as the availability of the starting materials and reagents, as well as the ease with which the subsequent functional group transformations can be made. However, it is important to note that these considerations are secondary to the establishment of the required carbon skeleton.

Consider now another skeletal arrangement and the 1,2-carbonyl addition reactions that might be used for its preparation. Benzylcyclohexane (**4**) is amenable to synthesis by addition of the benzyl group (or its equivalent) to the cyclohexane ring (or its equivalent).

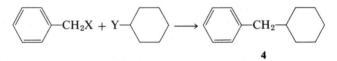

4

In practice, this synthetic plan may be achieved through an aldol-type condensation which will generate either 2-benzylidenecyclohexanone or 2-benzoylcyclohexanone.

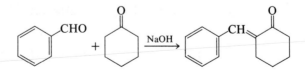

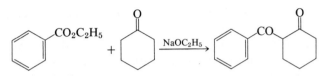

The same carbon skeleton is also available from the Grignard reaction between benzylmagnesium chloride and cyclohexanone, as well as from the Wittig reaction between benzyltriphenylphosphorane and cyclohexanone.

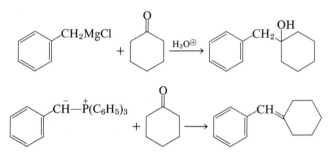

Additionally, the Grignard reaction may be employed in the reverse sense and still generate the desired carbon skeleton.

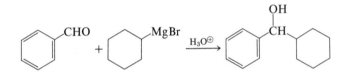

EXERCISE 4

Show how each of the foregoing products may be converted to benzylcyclohexane.

These examples by no means exhaust the various methods by which the structural arrangement of benzylcyclohexane (4) may be constructed. An important point is illustrated well, however, since the common denominator in the numerous methods shown is *the 1,2-addition of a nucleophilic species to a carbonyl group.* The substrates and reagents vary and depend on the reaction chosen, but the central feature of these carbon-to-carbon bond-forming reactions remains the carbonyl group. Due to the system where carbon is joined to a heteroatom by a multiple linkage, the carbon acquires a partial positive charge and is thus susceptible to union with another nucleophilic carbon species. This concept, then, unifies a large group of

organic reactions of slightly varying reaction mechanism into a more
synthetically useful grouping.

By expanding this concept to include substrates which have a double
bond in juxtaposition to the carbonyl group, we open the way to a new
variety of reaction types. Still available, of course, is the group of
1,2-carbonyl addition reactions; but now, because of the interaction of the
double bond and the carbonyl functions, a conjugate (1,4) addition of a
nucleophilic reagent is possible. The principle of the addition reaction is
still the same, but the mechanism varies a bit. Both acid (3) catalyzed and
base (4) catalyzed pathways are available, and which one is chosen depends
again on the substrates and reagents used.

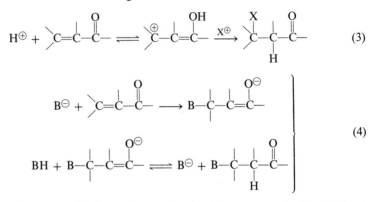

(3)

(4)

By far the most widely used reaction in this category is the Michael
reaction. By the proper choice of substrate and reagent, the carbon skeletons of both 3-methylhexane (3) and benzylcyclohexane (4) can be
prepared.

$$CH_3COC{=}CH_2 + CH_2(CO_2C_2H_5)_2 \xrightarrow{NaOC_2H_5} CH_3COCHCH_2CH(CO_2C_2H_5)_2$$
$$\quad\;\; \overset{|}{CH_3} \qquad\qquad\qquad\qquad\qquad\qquad\qquad\quad \overset{|}{CH_3}$$

$$CH_3CH{=}CHCOCH_3 + CH_2(CO_2C_2H_5)_2 \xrightarrow{NaOC_2H_5} (C_2H_5O_2C)_2CHCHCH_2COCH_3$$
$$\qquad\qquad\qquad\qquad\qquad\qquad\qquad\qquad\qquad\qquad\qquad\qquad\qquad \overset{|}{CH_3}$$

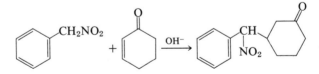

The conjugate addition of cyanide ion can also be used to prepare the
carbon skeleton of the 3-methylhexane (3)

$$CH_3CH_2CH{=}CHCOCH_3 \xrightarrow[DMF]{NaCN} CH_3CH_2CHCH_2COCH_3$$
$$\qquad\qquad\qquad\qquad\qquad\qquad\qquad\quad \overset{|}{CN}$$

and both 3-methylhexane (**3**) and benzylcyclohexane (**4**) are available by utilization of the appropriate Grignard reagent.

$$CH_3CH_2CH{=}CHCOCH_3 \xrightarrow[\substack{1)\ CuCl \\ 2)\ H_3O^{\oplus}}]{CH_3MgBr} CH_3CH_2\underset{\underset{CH_3}{|}}{CH}CH_2COCH_3$$

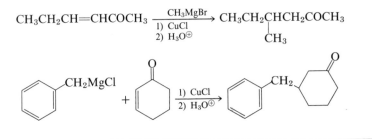

EXERCISE 5

(a) Show how each of the foregoing derivatives may be converted to 3-methylhexane and benzylcyclohexane, respectively.

(b) Cite an appropriate precedent for that sequence shown above that you would expect to be the most efficient route to each of these hydrocarbons.

It is important to remember that we are not elaborating the best method for the preparation of either of these hydrocarbons, but only that, through a common type of substrate, the carbon skeletons desired are available. Again, by grouping together several organic reactions that involve conjugate addition of a nucleophilic species to an unsaturated carbonyl system, we reveal another very general synthetic pathway.

Olefin addition reactions (type C)[2] also represent a rather general broad category of reactions which have similar synthetic usefulness. For instance, the carbon skeleton of 3-methylhexane (**3**) may be established by a Claisen rearrangement of the vinyl ether of 3-methyl-3-buten-2-ol

$$CH_3\underset{\underset{CH_3}{|}}{\overset{\overset{OCH{=}CH_2}{|}}{CH}}C{=}CH_2 \xrightarrow{\Delta} CH_3CH{=}\underset{\underset{CH_3}{|}}{C}CH_2CH_2CHO$$

and that of benzylcyclohexane (**4**) prepared through the use of the Diels–Alder reaction below.

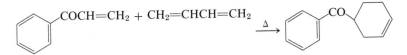

---

[2] See C. DePuy, *Molecular Reactions and Photochemistry*, and L. Stock, *Aromatic Substitution Reactions*, in Foundations of Modern Organic Chemistry Series, Prentice-Hall, Inc., Englewood Cliffs, N.J., 1968 for a more detailed discussion of these type C reactions.

Similarly, both carbon skeletons are amenable to synthesis by application of the Friedel–Crafts reaction when the appropriate substrates and reagents are employed.

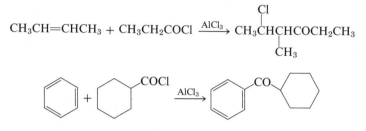

$$CH_3CH{=}CHCH_3 + CH_3CH_2COCl \xrightarrow{AlCl_3} \underset{\underset{CH_3}{|}}{CH_3}\overset{\overset{Cl}{|}}{C}HCHCOCH_2CH_3$$

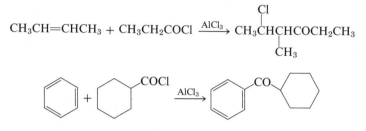

The unifying feature of this group of reactions is found in the interaction of a reagent with a carbon-to-carbon double bond. Mechanistic features vary from the electron reorganization types, such as the Diels–Alder reaction and the Claisen–Cope rearrangement, to the polar types, such as the Friedel–Crafts reaction. Inherent in both, however, is the underlying importance of the double bond. This multiple linkage provides the site for the formation of a new carbon-to-carbon bond; hence, it provides for the construction of new carbon systems.

Finally, another class of organic reactions is evident in the displacement reactions.[3] Most common among this group is the alkylation reaction, and the following examples emphasize its use.

$$CH_3CH_2COCH_2CO_2C_2H_5 + C_2H_5I \xrightarrow[C_6H_6]{NaH} CH_3CH_2COCHCH_2CH_3$$

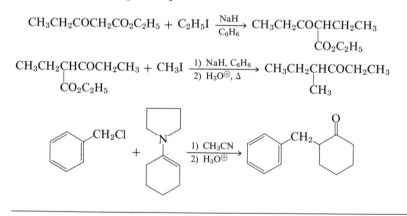

EXERCISE 6

Characteristic of the formation of a carbon-to-carbon bond by the displacement reaction is the generation of a carbanionic species. An "activating group" is

---

[3] These type D reactions are discussed in detail by W. H. Saunders, Jr., *Ionic Aliphatic Reactions* (1965), and C. D. Gutsche, *The Chemistry of Carbonyl Compounds* (1967) in Foundations of Modern Organic Chemistry Series, Prentice-Hall, Inc., Englewood Cliffs, N.J., and the reader is directed to these works for review.

commonly added to facilitate this process. In the following transformations, suggest suitable activating groups.

(a) $CH_3COCH_2CO_2C_2H_5$ and $CH_2{=}CHCH_2Br \longrightarrow CH_3COCH_2CH_2CH{=}CH_2$

(b) $(CH_3)_2CHCOCH(CH_3)CHO$ and $CH_3I \longrightarrow (CH_3)_2CHCO\overset{\overset{\displaystyle CH_3}{|}}{C}HCH_3$

(c) $C_6H_5CH(CO_2C_2H_5)_2$ and $CH_3CH_2I \longrightarrow C_6H_5\overset{\overset{\displaystyle C_2H_5}{|}}{C}HCO_2H$

(d) $CH_3CH{=}CH{-}N\;\square$ and $C_6H_5CH_2Br \longrightarrow CH_3CH_2\overset{\overset{\displaystyle CHO}{|}}{C}HCH_2C_6H_5$

(e) $CH_3CH(CO_2C_2H_5)_2$ and $\overset{O}{\overset{\triangle}{CH_2{-}CH_2}} \longrightarrow CH_3\text{-}\square$

The basic mechanistic features of the reactions in this group are well understood and involve the interaction of a nucleophilic species with a more or less positive carbon atom. Rather than employing a multiply bonded heteroatom to impart positive character on the reactive carbon as in the carbonyl group, we rely here on the polar character of a carbon-heteroatom single bond. In general, this requires a more nucleophilic species in order to observe reaction. The synthetic result, however, is similar in that two carbon units are joined by a carbon-to-carbon bond. Thus the difference in the two reaction types A and D lies principally in the substrate used, and their separation into types is useful when thinking in synthetic terms.

EXERCISE 7

Using as the carbon-to-carbon bond-forming reaction a representative of the reaction type indicated, show how the carbon skeletons of the following compounds can be prepared from the starting materials shown.

(a) $CH_3CH_2COC_6H_5 \xrightarrow{\text{type A}} CH_3CH_2\overset{\overset{\displaystyle C_6H_5}{|}}{CHCH_2CO_2H}$

(b)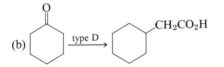

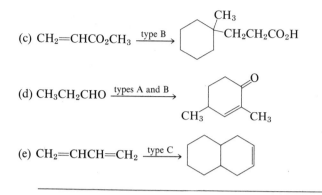

(c) $CH_2{=}CHCO_2CH_3 \xrightarrow{\text{type B}}$

(d) $CH_3CH_2CHO \xrightarrow{\text{types A and B}}$

(e) $CH_2{=}CHCH{=}CH_2 \xrightarrow{\text{type C}}$

Again, the foregoing classification of the carbon-to-carbon bond-forming reactions is useful in *organizing* the tools of synthesis. It is obviously a rough order of things, for the lines of demarcation are not clear-cut. The variations in reaction type are subtle and provide a continuum from one type to another. The Diels–Alder reaction between butadiene and phenyl vinyl ketone can be considered a conjugate addition reaction as well as an olefin addition reaction, and the carbenoid reagents appear in virtually all categories. However crude the classification may be, it has merit in its utility as long as the system is not taken with too great a degree of rigidity. Recognize that we are speaking here of *carbon-to-carbon bond-forming types that serve to prepare various carbon skeletons.* Within this limitation, these generalizations are useful, for they make easier the task of systematizing the synthetic process. To synthesize any compound from its less complex components, we must first be able to construct the carbon framework.

Scrutiny of the various derivatives of the 3-methylhexane (3) and benzylcyclohexane (4) systems that we have prepared reveals quite a variety of functionally substituted systems. This feature of the various organic reactions available in synthesis will become important when we consider the synthesis of more specific compounds rather than just carbon skeletons. In particular, one specifically desired functional array may preclude the use of one or more reaction types in its synthesis. However, there is variation possible among the functional groups, and it will probably be more profitable to modify functional groups than to switch carbon-to-carbon bonding-forming reaction types. In this connection, it is worth mentioning some of the more important functional group modifications available.

Again an incomplete list of types is presented for orientation and review. Now we have classified all the reactions of organic chemistry somewhat crudely, but with the emphasis on their synthetic utility.

FUNCTIONAL GROUP MODIFICATIONS[4]

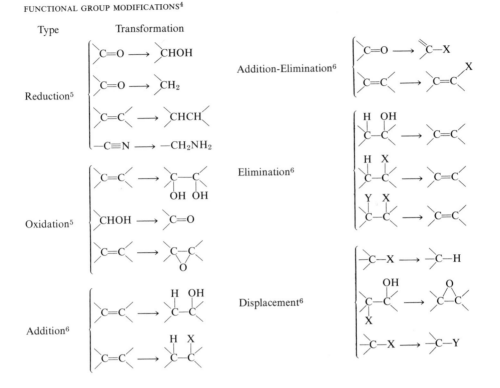

Consider first how these changes in functional group are useful in organic synthesis. Were it the goal of a synthesis to prepare 2-methylpropylbenzene (**5**), we would need to use intermediates which contained some functional group in order to make the carbon skeleton.

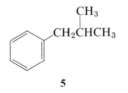

5

[4] For a general discussion of functional groups, see O. L. Chapman, *Functional Groups in Organic Compounds,* in Foundations of Modern Organic Chemistry Series, Prentice-Hall, Inc., Englewood Cliffs, N.J., 1968.

[5] For detailed coverage of these reactions, see K. L. Rinehart, Jr., *Oxidation and Reduction of Organic Compounds,* in Foundations of Modern Organic Chemistry Series, Prentice-Hall, Inc., Englewood Cliffs, N.J., 1968.

[6] The intricacies of these transformations are covered by R. Stewart, *The Investigation of Organic Reactions* (1966), and W. H. Saunders, Jr., *Ionic Aliphatic Reactions* (1965), in Foundations of Modern Organic Chemistry Series, Prentice-Hall, Inc., Englewood Cliffs, N.J.

Thus some reactive site is necessary to unite the carbon atoms, but we must be prepared to alter these reactive centers so as to obtain the final product. We can choose several alternative pathways which lead to the hydrocarbon **5**; three of these are represented below.

Carbon-to-carbon
bond-forming reaction   Functional group modification

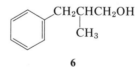

Notice that while, in each case, the same hydrocarbon **5** is obtained, the differing mode of carbon-to-carbon bond-forming step dictates the use of a different set of conditions to modify the resulting functional group. These two phases of any synthetic effort are mutually dependent on one another, but the sequence of events remains the same: formation of the carbon skeleton and then manipulation of functional groups.

Let's now change the desired end product slightly and consider the synthesis of 3-phenyl-2-methylpropanol (**6**). In this case we must leave a

CH₂CHCH₂OH
|
CH₃

6

functional group in the molecule we construct. Such a situation places even more restrictions on the mode of carbon skeleton synthesis we choose. We must now arrange to unite the necessary carbon atoms through two reactive centers and also prepare the resulting molecule for inclusion of the alcohol group on the desired carbon atom. Our primary concern, then, will be to search the carbon-to-carbon bonding reaction types for methods that leave at least a precursor of the alcohol function in the molecule. An obvious candidate that meets these requirements is a reaction of the type B, and a possible synthesis is shown below.

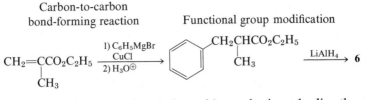

To an even greater degree than before, this synthesis embodies the close interplay of carbon skeleton construction and subsequent functional group transformations. There are certainly other ways we can construct the alcohol **6**, but in each case, the reasoning would be the same: "If we use reaction type ___ to make the skeleton, then we can modify the functional group by method _____."

EXERCISE 8

Show the reactions that are necessary to carry out the following transformations, and indicate their types.

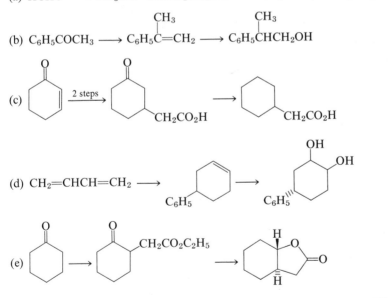

The aim of the discussion so far has been to bring out the differences in two major groups of organic reactions in preparation for their application in synthetic planning. While neither major classification is independent of the other in their use, there is more variation possible in the manipulation of functional groups than in the formation of the carbon skeleton. The reagents available for the formation of carbon-to-carbon bonds are rather limited, for instance, compared to those available for the reduction of carbonyl groups. For the latter specialized conversion, we may choose from various aluminum and boron hydrides, hydrogen, and numerous noble metal catalysts, dissolving metals in protic solvents and metal alkoxide-ketone couples, to mention a few. The conditions of temperature, pressure, and reaction time all greatly affect the ease and selectivity of such reductions. Synthetic planning, then, is a balance between the problem of framework construction through the use of carbon-to-carbon bond-forming reactions and the problem of subsequent functional group manipulations. The importance of each will vary through rather narrow limits, but the presence of the two major classes will always be evident.

REFERENCES

In addition to the appropriate books in this series, the following works present a more comprehensive study of the organic reaction types cited in this chapter.

H. O. House, *Modern Synthetic Reactions* (New York, N.Y.: W. A. Benjamin, Inc., 1965); R. B. Wagner and H. D. Zook, *Synthetic Organic Chemistry* (New York, N.Y.: John Wiley & Sons, Inc., 1953); L. F. Fieser and M. Fieser, *Reagents for Organic Synthesis* (New York, N.Y.: John Wiley & Sons, Inc., 1967); R. A. Raphael, *Acetylenic Compounds in Organic Synthesis* (London, England: Butterworth and Co., Ltd., 1955); D. C. Ayres, *Carbanions in Synthesis* (London, England: Oldbourne Press, 1966); C. W. Bird, *Transition Metal Intermediates in Organic Synthesis* (London, England: Logos Press-Academic Press, 1967).

An in-depth survey of a limited number of these reactions is contained in the series, *Organic Reactions* (New York, N.Y.: John Wiley & Sons, Inc.), and specific experimental conditions for carrying out many of these reactions are contained in *Organic Synthesis* (New York, N.Y.: John Wiley & Sons, Inc.).

2

Synthetic Design I
Preliminary Planning

If there is any key to success in planning a synthesis, it is to work the problem backwards. This is really the cardinal rule of synthesis. Beginning with the total concept of the molecule desired and all of its structural ramifications, we methodically break it apart, piece by piece, in such a way that we can best predict success in reassembling the pieces. We tackle each problem as it is presented and work toward solutions that will leave the resulting synthetic tasks simpler than the ones just solved. When we have successfully unraveled the complex molecular network and proposed how each component may be rejoined in its turn to reconstruct our objective, the synthetic plan is in hand. In designing a synthesis, we must think back from the complex to the simple so that, in practice, we may rationally work from the simple to the complex with a suitable map to follow. It is impossible to overemphasize this concept, and our further discussion of synthetic planning is predicated on demonstrating its importance.

MOLECULAR HISTORY

Suitable synthetic objectives come in all sizes, shapes, and orders of difficulty. Before beginning any concrete effort at planning a synthesis, one should consider the knowns and unknowns of the desired system. The results of previous investigations often have a very important bearing on the scheme chosen. Particularly important are the results obtained by other investigators who may have attempted a synthesis earlier. If their efforts were unsuccessful, the reasons for their failure must obviously be noted. For instance, in an approach to the synthesis of dehydroabietic acid (**9**), the Reformatsky reaction on the ketoester **7** failed[1] to generate any of the expected hydroxyester **8**. For anyone considering the synthesis of dehydroabietic acid (**9**), this information is indeed important. A successful synthetic scheme for this molecule will have to modify or obviate this particular stage.

---

[1] W. E. Bachmann, G. I. Fujimoto, and L. B. Wick, *J. Am. Chem. Soc.,* **72,** 1955 (1950); W. E. Bachmann and L. B. Wick, *ibid.,* **72,** 2000 (1950).

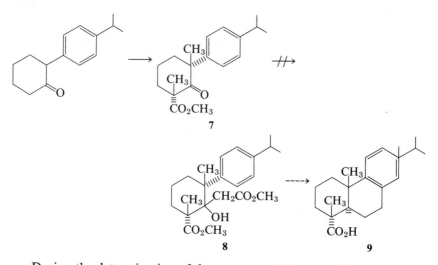

7

8   9

During the determination of the structure of many natural substances, particular characteristics of the molecule usually come to light. This information is very useful in synthetic work, for we learn how the molecule's framework affects the character of the functionality present. For instance, the degradative work that led to the structure of cortisone (**10**) indicated that the hindered C-11 ketone was nonreactive toward normal

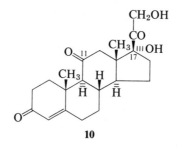

10

ketonic reagents such as the hydrazines and that the C-17 dihydroxyacetone side was quite labile, particularly toward oxidation. Both of these pieces of information proved very useful in synthetic projects directed toward cortisone (**10**). Partial syntheses took advantage of the inert character of a C-11 oxygen substituent to elaborate the labile C-17 side chain as the final stages, while total syntheses had to cope with the introduction of the oxygen function at the unreactive and inaccessible C-11 position. The degradative work, however, suggested that, once introduced, a C-11 substituent could be readily handled in the presence of other similar substituents due to its hindered character.

In some instances, during the degradative work to determine the structure of natural materials, partial syntheses will be accomplished. Thus chemical degradation may generate a substance from which reconstruction of the natural product can be readily accomplished. In the degradative work, such partial syntheses serve to establish part structures; in total synthetic work, such partial syntheses simplify the problem to one of construction of the degradative intermediate only. A case in point is found in the chemistry of the terpene camphor (**11**).[2]

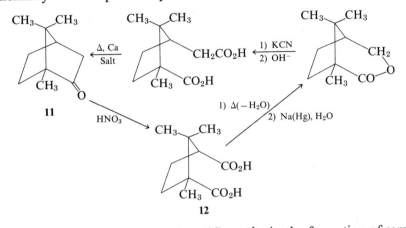

Oxidative degradation of camphor (**11**) results in the formation of camphoric acid (**12**). To establish the structural change that attends this degradation, camphoric acid (**12**) was reconverted to camphor (**11**) by the process shown above. With this information in hand, the total synthesis of camphor entails only the synthesis of camphoric acid (**12**). This approach to total synthesis is referred to as the *relay approach*. In this method, the natural substance may not itself be totally synthesized; rather, a rational pathway from commercial starting materials to the natural product is established by use of intermediate degradation products. Thus, a synthesis of a key intermediate degradation product is accomplished; the racemic synthetic material is resolved; its identity with the natural material is established; and finally, the synthesis of the natural product itself is completed using material obtained by degradation rather than synthesis.

Another example of the *relay approach* to total synthesis is the construction of phyllocladene (**15**)[3] shown below. The ketoacid **14** was a key substance in the degradative proof of the structure of phyllocladene (**15**) and was readily available from the natural product. This ketoacid **14** proved to be an ideal intermediate in the total synthesis of phyllocladene (**15**). By using ketoacid **14** obtained from degradation of phyllocladene, Turner and his co-workers were able to replenish meager supplies of the intermediate that resulted from synthesis. The ketoacid **14** was synthesized

---

[2] The chemical history of this valuable terpene is covered by J. L. Simonsen, *The Terpenes*, Vol. 2 (London: Cambridge University Press, 1949), pp. 373–383.

[3] R. B. Turner, K. H. Gänshirt, P. E. Shaw, and J. D. Tauber, *J. Am. Chem. Soc.*, **88**, 1776 (1966).

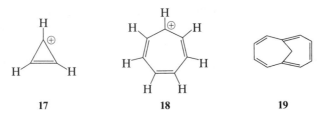

17   18   19

among others should exhibit this stability. Important to the synthetic effort in these cases is that these molecules are predicted to be more stable than their polyene structures would suggest. Therefore synthetic methods employed in their construction need not be delicate. This does not mean that the syntheses must be uninspired. The presumed stability of the products does, however, suggest that experimental conditions need not be exceedingly gentle.

In contrast to this situation, an equally valid test of the Hückel rule is the synthesis of molecules predicted to be nonaromatic and hence polyenic. Molecules such as pentalene (**20**) and cyclobutadiene (**21**) should gain no

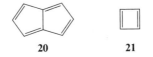

20   21

stabilization from electron delocalization and indeed are predicted to be particularly unstable. This prediction must be borne in mind in planning any synthetic approach to these systems. Very delicate operations should be foreseen, and sequences should be chosen on the basis of the presumed instability of the objective. When the inability to synthesize a molecule is to be taken as justification for a theory, it is absolutely necessary to establish that the negative results pertain to the molecular structure and not to the synthetic sequence or experimental conditions. This may be a hard task, and it is usually better to seek positive evidence through synthesis, rather than negative. In the preceding two cases, both molecules have been synthesized but not in the free forms indicated. Recognition of the results of molecular orbital calculations which suggest electron deficiencies in the

$Fe(CO)_3$

22   23

parent molecules **20** and **21** led to construction of the systems **22**[9] and **23**,[10] which again provide substantiation of theory.

[9] T. J. Katz, M. Rosenberger, and R. K. O'Hara, *J. Am. Chem. Soc.,* **86,** 249 (1964).
[10] G. F. Emerson, L. Watts, and R. Pettit, *ibid.,* **87,** 131 (1965).

KEY INTERMEDIATES

Throughout the foregoing discussion, we emphasized that the start of any synthetic design begins with a survey of the objective in the broadest sense and then works toward the particular. Thus we have begun to work our synthesis backwards by first evaluating the molecule in the framework of existing knowledge. We turn now to the somewhat closer scrutiny of the synthetic objective to determine the gross features of the project. We may find it advantageous to subdivide the synthesis into smaller components for the purpose of more synthetic latitude and/or more efficient construction. This process is one of recognizing key structural subunits through which a synthetic scheme may pass. It is akin to finding the high spots in the terrain to be traversed and then focusing attention on each as it comes up. This is particularly useful when extended syntheses of large molecules are contemplated. We select key intermediates along the way that are themselves challenging efforts and which, when synthesized, will serve as starting materials for the next stage. Two examples of this approach to planning were mentioned earlier in the camphor (**17**) and phyllocladene (**15**) syntheses. Here the key intermediates were suggested from degradative work, and their availability from the natural materials automatically suggests their utility in the synthetic scheme. In the case of the phyllocladene (**15**) synthesis, the effect of choosing the ketoacid **14** as a key intermediate is to subdivide the total synthesis into two projects —one concerned with the synthesis of the ketoacid **14** itself, and the other with its conversion to the desired end product. Thus, before any actual synthetic planning is done, the choice of the ketoacid **14** as a key intermediate has circumscribed the approach and represents a high spot along the way.

Key intermediates may evolve through yet another evaluation of the synthetic project. In many instances, variation in molecular structure occurs in only one portion of the molecule. For example, several resin acids differ from one another in the substitution pattern found in one ring only, and the remainder of the molecule is the same throughout. In cases such as these, a key intermediate is suggested by consideration of the latitude desirable in a synthetic scheme. Inasmuch as the first two rings of all of these acids have the same structure, it would be particularly reward-

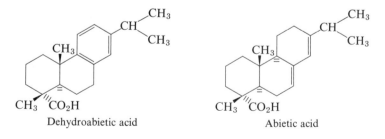

Dehydroabietic acid   Abietic acid

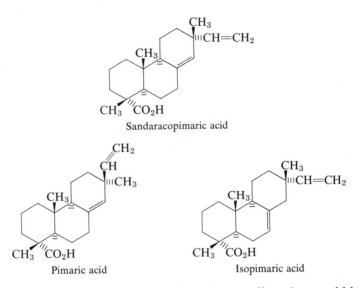

Sandaracopimaric acid

Pimaric acid   Isopimaric acid

ing to construct a suitably substituted key intermediate that would lead to all. Thus, without the necessity of degradative work, we can schematically suggest that the aromatic acid **24** might serve as a common precursor to

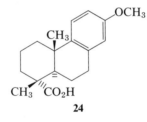

24

all the above resin acids. Again we have circumscribed our synthetic effort by settling on a high spot along the way. This time the key intermediate is chosen because of the latitude it affords for the synthesis rather than for its usefulness as a relay compound. When general syntheses of a class of compounds are contemplated, the structures of the various members of the class may indeed suggest this key intermediate approach. Again we see that the synthetic planning process is facilitated by a concern for the final objective first. The ability to designate key intermediates or high spots in a prospective synthetic effort easily and rapidly provides guidelines for the effort. These may evolve from prior knowledge of the particular field, or from careful examination of the molecular species involved; they constitute the first rudiments of a synthetic plan.

STARTING MATERIALS

Before we begin the actual detail of synthetic design, we must define the starting point as well as the final objective. The planning process involves reducing the complex end product back to simple starting materials. But where is this uncomplex objective? Theoretically, any synthesis can be reduced to coal, air, and water as the initial ingredients; however, this is absurd in practice. What, then, may we consider starting material?

Starting materials are any readily available substances that can be obtained routinely and in quantity. This usually implies commercially available materials which can be purchased from the fine-chemical houses, as well as the bulk materials which have a wider distribution. These are the compounds with which one actually begins laboratory work, and in practice, a synthetic scheme must be reduced to these terms. The student of organic synthesis should be familiar with the compounds and types of compounds on the market in order that he have a feeling for where his synthetic planning may end. In general, commercial materials are rather simple compounds with wide appeal to the organic chemist. For instance, methyl iodide, cyclohexanone, and cinnamic acid are readily and inexpensively available from almost any supply house because they satisfy many chemists in many diverse endeavors. However, a vast array of much more exotic compounds is also available. They usually cost more because the market is smaller; but when needed, the effort saved may justify their purchase. For example, many steroids and heterocyclic compounds are manufactured commercially; so are molecules like

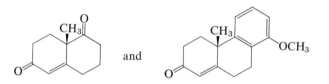

and

The only sure way to be familiar with the commercially produced molecules is to peruse the catalogs of the fine-chemical companies. Their catalogs are revised frequently, and continual awareness of catalog changes is worthwhile. The references at the end of this chapter include a partial list of the larger fine-chemical companies whose catalogs are instructive.

While laboratory work must begin with materials that may be purchased, research in synthesis does not necessarily begin there. There is a large domain of materials that are not commercially available but are so well documented that the substances may be prepared routinely and in quantity by means of described laboratory procedures. Although little or no research is necessary to construct these compounds, their general utility to the scientific community is not great enough to warrant their production on a commercial scale. The procedures for the preparation of such compounds are found in the original literature and *Organic Syntheses*. The latter compilation is a particularly important source of synthetic intermediates. Collected into four volumes, each representing the compilation of ten yearly volumes, *Organic Syntheses* is educational reading as well as a gold mine of materials and techniques. Before any synthesis is begun, it is worthwhile checking this work to ascertain whether a starting material or pertinent procedure is recorded therein. Inasmuch as all the *Organic Syntheses* procedures have been subjected to an independent check by chemists other than the originators of the procedure, a user may be assured that he will obtain the desired material.

Finally, it must be recognized that there is a time lag before standard preparations of potentially useful starting materials find their way into commercial catalogs or compendiums like *Organic Syntheses*. Before this can happen, the procedures must have been thoroughly investigated and must be so routine that experimental success is assured. In the interim, these materials may still be prepared with little effort by the research procedures recorded in the chemical literature. Therefore, the research literature is the final source of procedures which may be used to make starting materials. Though these procedures will not have been double-checked as in *Organic Syntheses,* they still may be just as reliable. In any event, the record of a specific compound having been prepared can serve to justify its use in a projected synthesis.

Starting material for synthetic work, then, falls into two categories. The compounds with which we actually begin work in the laboratory are those which are commercially available. Materials that are routinely prepared in quantity by means of well-documented procedures also are readily available in the laboratory and may be treated as starting materials from a synthetic planning point of view. The former are the compounds with which we will actually begin our laboratory work; the latter are the group of materials to which we endeavor to reduce the complex, desired end products in the design phase of synthesis.

REFERENCES

Fine-chemical suppliers:

  1. Aldrich Chemical Co., 2371 North 30th St., Milwaukee, Wis. 53210.
  2. J. T. Baker Chemical Co., Phillipsburg, N.J. 08865.
  3. Eastman Organic Chemicals Dept., Rochester, N.Y. 14603.
  4. Fisher Scientific Co., 717 Forbes Ave., Pittsburgh, Pa. 15219.
  5. Matheson, Coleman, and Bell, 2909 Highland Ave., Norwood (Cincinnati), Ohio 45212.

For a more comprehensive list, see L. F. Fieser and M. Fieser, *Reagents for Organic Synthesis* (New York, N.Y.: John Wiley & Sons, Inc., 1967).

3
Synthetic Design II Molecular Characteristics

INTRODUCTION

The heart of any synthetic effort is *the plan*. R. B. Woodward[1] makes this point very emphatically when he writes: "synthetic objectives are seldom, if ever, taken by chance, nor will the most painstaking or inspired, purely observational activities suffice. Synthesis must always be carried out by plan, and the synthetic frontier can be defined only in terms of the degree to which realistic planning is possible, utilizing all of the intellectual and physical tools available." The success of a synthetic program depends upon both the chemist's intellectual prowess in planning the effort and the maturity of the field. More than any other scientific endeavor, a synthetic plan is an effort on the part of the organic chemist to put his knowledge of the field to his own use. He must project the results of experiments as yet untried and unite numerous such projections into a cohesive scheme that will accomplish his preconceived goal. We look now to some of the concepts that make up the synthetic plan.

MOLECULAR CHARACTERISTICS

A. MOLECULAR SIZE

At the outset of any synthesis, the ultimate objective is firm in mind. It is well, at this early stage, to take the measure of our adversary and evaluate some of the characteristics that will be important to the synthesis.

The magnitude of any project will depend in some measure on how large a molecule is desired. On the basis of molecular size alone, the problems inherent in the synthesis of 2-methyl-3-phenylpropanol (6) are not nearly those associated with a steroid synthesis. Thus a consideration of

---

[1] R. B. Woodward, "Synthesis" in *Perspectives in Organic Chemistry* (New York, N.Y.: Interscience Publishers, Inc., 1956), p. 155.

the number of atoms which must be united brings up the problem of logistics. In general, the larger the molecule desired, the greater the number of steps required to accomplish the synthesis. The result is that the

$$CH_2{=}\overset{\underset{\displaystyle |}{CH_3}}{C}CO_2C_2H_5 \xrightarrow[\substack{2)\ H_3O^+ \\ 75\%}]{1)\ C_6H_5MgBr \\ CuCl} C_6H_5CH_2\overset{\underset{\displaystyle |}{CH_3}}{C}HCO_2C_2H_5 \xrightarrow[90\%]{LiAlH_4}$$

$$C_6H_5CH_2\overset{\underset{\displaystyle |}{CH_3}}{C}HCH_2OH$$
6, 68%
overall

arithmetic demon raises his ugly head and works to thwart success. The two-step synthesis of 2-methyl-3-phenylpropanol (**6**) may be accomplished in 68% overall yield from commercially available materials, while one of the best total syntheses of the racemic steroid estrone (**26**) requires seven steps and results in only an 8.5% overall yield from commercially available material.

It is, of course, obvious that a steroid synthesis is several orders of magnitude more complex in many ways than a synthesis of the propanol derivative **6,** but the effect on the overall yield of having to employ more reactions to synthesize larger molecules is worth emphasis. A sequence of

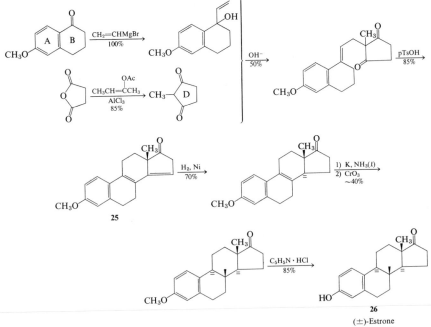

25

26
(±)-Estrone
8.5% overall

seven consecutive steps, each of which affords a 90% yield, will result in a $(0.90)^7 \times 100 = 47\%$ overall yield. It is rare when seven organic reactions that each afford a 90% yield can be strung together sequentially; therefore, we can expect less than a 47% overall yield in practice. Logistics in synthesis is a very real factor with which we must be concerned.

The arithmetic demon dictates one of the major axioms of synthesis: "Get the most done in the fewest steps and in the highest yield." In planning a synthesis, care must be taken that the most efficient reaction sequence is chosen to overcome a given structural feature. We have already seen that there are numerous ways to construct the carbon skeletons of such molecules as 3-methylhexane (3) and benzylcyclohexane (4). Before a synthetic plan for either of these compounds is complete, it would be necessary to evaluate each set of reactions proposed and try to ascertain which set would result in the highest yield. Several factors enter into this decision, and many subtle influences are involved. However, three major considerations are important: the number of steps, the availability of starting material, and the type of reactions in each proposed scheme. We will consider the latter two criteria later, and need only reiterate that the fewest operations possible make the best number of steps for a synthesis.

At the planning stage, there is one efficient way to reduce the sequential number of steps required—namely, do as much of the work in parallel as possible. It is far more efficient to work separately on preparing two large chunks of a molecule and then to add them together at the latest possible stage than to put the entire molecule together piece onto piece. Consider a seven-step synthesis (A $\longrightarrow$ H) done first in a linear[2] fashion and then in a convergent[2] manner. If each reaction afforded a 90% yield, the linear

$$A \xrightarrow{90\%} B \xrightarrow{90\%} C \xrightarrow{90\%} D \xrightarrow{90\%} E \xrightarrow{90\%} F \xrightarrow{90\%} G \xrightarrow{90\%} H \qquad 47\% \text{ overall}$$

$$\begin{array}{c} A \xrightarrow{90\%} B \xrightarrow{90\%} C \\ (81\%) \\ \\ D \xrightarrow{90\%} E \xrightarrow{90\%} F \\ (81\%) \end{array} \Biggr\rangle \xrightarrow{90\%} G \xrightarrow{90\%} H \qquad 64\% \text{ overall}$$

process would result in a 47% yield of H, while the convergent scheme would produce a 64% yield of H. In reality, the convergent approach can be considered a four-step synthesis instead of seven.

The mechanics of the convergent scheme also augur in its favor. In this approach, it is easier to build up supplies of intermediate materials, as work is always closer to starting material. The linear approach requires the continual adding of new fragments onto a recently prepared intermediate.

[2] These descriptive terms for these general approaches were first proposed by L. Velluz, J. Ralls, and G. Nominé, *Angew. Chem., Int'l. Ed.*, 4, 181 (1965).

An example of a convergent synthesis is the ($\pm$) estrone (**26**) synthesis outlined earlier. In this efficient scheme of the steroid, the A/B ring system is joined to a preformed D ring through a two-carbon bridge. The yield (based on the lower-yield branch of the parallel system) of the tetracyclic molecule **25** that possesses the required estrone carbon skeleton is quite high (36%) when one considers the size of the molecule.

Therefore, if there is a choice between two schemes which differ in this mode of uniting intermediates, the arithmetic demon will be most foiled by the convergent synthesis. Of course, other considerations will enter into the decision as well, but we should never lose sight of this basic solution to the overall yield problem.

B. CARBON SKELETAL COMPLEXITY

Not only will the size of the synthetic objective be an important factor at the planning stage, but the complexity of the carbon framework also must be considered. Some very small molecules present significant synthetic obstacles solely because of the unusual carbon-to-carbon bonding. Consider the hydrocarbon cubane (**27**). This molecule contains only eight carbon atoms; but as a result of the unusual cyclobutane structure, the reported synthesis requires fifteen steps.[3] Here the problem is not one of putting large numbers of carbon atoms together, but rather, one of devising a series of reactions that will constrain a few carbons in the

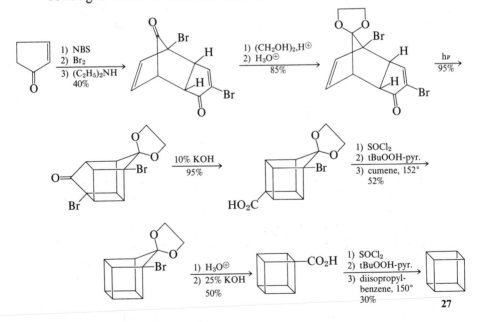

[3] P. E. Eaton and T. W. Cole, Jr., *J. Am. Chem. Soc.*, **86**, 3157 (1964).

desired arrangement. Thus, carbon skeletal complexity severely restricts the breadth of reaction choice, and to gain the necessary selectivity, a disproportionate number of stages is required compared to the molecular size. Notice that the much larger estrone molecule (**26**) can be constructed in seven steps, while the presently known synthesis of cubane (**27**) requires fifteen. Hence, another facet that must be seriously considered in planning any synthesis is the complexity of the desired carbon skeleton.

Through the efforts of numerous chemists who were faced with carbon skeletal problems of varying complexity, organic chemistry is blessed with many efficient sequences of reactions which overcome some of the more common skeletal problems. For instance, the advent of preparative photochemistry has made the cyclobutane ring system much more readily achieved. The foregoing cubane synthesis is an example of this. Recognition of the ease with which four-membered rings can be obtained photochemically suggests the use of this method for the introduction of this arrangement in the synthesis of such complex molecules as caryophyllene[4] (**28**) and annotinine[5] (**29**). Indeed, this structural feature of both of these molecules was found amenable to construction by photochemical means.

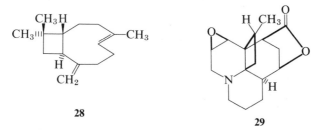

28

29

Work on the total synthesis of the steroid skeleton very early suggested that an efficient method of fusing one six-membered ring to another would be invaluable to the synthetic effort. A combination of the Mannich, the Michael, and the aldol reactions—together known as the Robinson

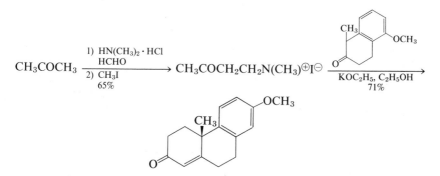

$$CH_3COCH_3 \xrightarrow[\substack{2)\ CH_3I \\ 65\%}]{\substack{1)\ HN(CH_3)_2 \cdot HCl \\ HCHO}} CH_3COCH_2CH_2N(CH_3)^{\oplus}I^{\ominus} \xrightarrow[\substack{KOC_2H_5,\ C_2H_5OH \\ 71\%}]{}$$

[4] E. J. Corey, R. B. Mitra, and H. Uda, *J. Am. Chem. Soc.,* **86,** 485 (1964).
[5] E. H. W. Böhme, Z. Valenta, and K. Wiesner, *Tetrahedron Letters,* 2441 (1965).

annelation sequence—provided a powerful answer to this skeletal feature. Since this process was initially described, it has undergone careful scrutiny and modification, and the sequence has taken its place as a standard method of such ring constructions. Thus, consideration of the synthesis of such diverse molecules as cortisone (**30**) and emetine (**31**) has suggested the use of the Robinson annelation sequence for the six-membered rings in bold face.

One of the most powerful organic reactions is the Diels–Alder condensation; it has become the method of choice for the construction of bridged

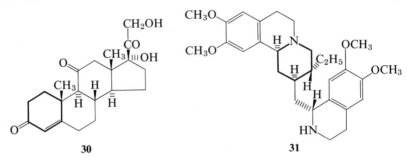

30   31

carbocyclic systems, as well as more standard nonbridged six-membered ring skeletons. Here, should the desired synthetic objective possess a similar

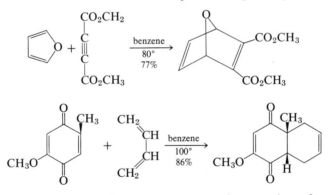

bicyclic ring system, even though it may be only a portion of a complex molecule, the Diels–Alder reaction with appropriately substituted reactants may provide the answer to the carbon skeletal complexity. Examples of substances which have yielded to synthesis through the use of this tool are numerous and span molecular size from cantharadin (**32**)[6] to cholesterol (**33**)[7] and triptycene.[8]

[6] G. Stork, E. E. van Tamelen, L. J. Friedman, and A. Burgstahler, *J. Am. Chem. Soc.*, **75**, 384 (1953).

[7] R. B. Woodward, F. Sondheimer, D. Taub, K. Heusler, and W. M. McLamore, *J. Am. Chem. Soc.*, **74**, 4223 (1952).

[8] G. Wittig, *Org. Syn. Coll.*, Vol. IV, 964 (1963).

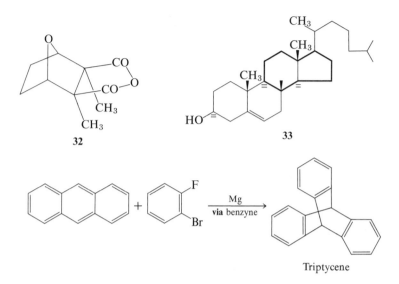

32

33

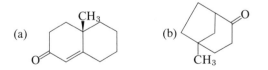

Triptycene

In the foregoing discussion, it is important to recognize not so much the reaction sequences cited as the approach to molecular construction. There are far too many sequences available for the synthesis of specific carbon skeletal arrangements to do justice in any kind of brief discussion. However, what does emerge is an approach one can take toward solving the complexities of carbon arrangements. Certain features of most molecules have analogies in less complex systems for which a synthetic method exists. If in planning a synthesis we concentrate less on the individual carbons and more on the portions of the molecule that form distinct subunits, reaction sequences of the type just discussed will become more obvious. In essence, a synthesis of a complex carbon skeleton must not be approached piecemeal, but as the union of various sections. In this manner, many of the most complex skeletal arrangements become much simpler and amenable to rational synthetic design.

EXERCISE 9

Outline the portion(s) of the following molecules that represent potential synthetic subunits.

(a)

(b)

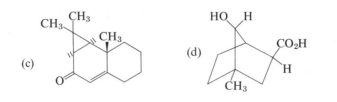

(c)

(d)

C. FUNCTIONALITY

Another important consideration of any synthetic program is the character of the functional groups present in the desired final product. While we have emphasized the importance of the skeletal features of organic molecules, we cannot overlook the problems presented by diverse functionality. It is true that, before any molecule can be made, a route to the carbon skeleton is absolutely necessary; however, no molecule can be completely constructed without elaboration of the functional group array that is present. Therefore, any scheme for overcoming the logistic problems of large molecules and/or the complexities of skeletal arrangements will necessarily be tempered by the resultant functionality desired.

Many sets of functional groups present little or no problems. For instance, alcohols may be readily prepared from aldehydes, ketones, or esters, and the choice depends on which grouping is more valuable in the intermediate synthetic stages. We have encountered this situation already in the synthesis of 2-methyl-3-phenylpropanol (6) where an ester function was necessary for the skeletal construction and was readily converted to the desired alcohol by reduction. An equally routine conversion is presented by the synthesis of the diterpenoid hydrocarbon (±) kaurene (35).[9] The conversion of the ketonic intermediate (34) into the desired olefin 35 is a routine matter through application of the Wittig reaction.

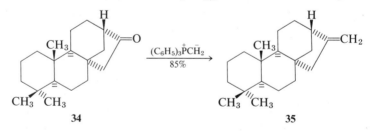

34   35

The complex carbon skeleton of this system is virtually inert and presents no difficulty to this final functional group transformation. The greater utility of the ketone 34 over the olefin 35 in previous synthetic stages is apparent from the previous discussion of carbon-to-carbon bond-forming reactions.

[9] R. A. Bell, R. E. Ireland, and R. A. Partyka, *J. Org. Chem.,* **31,** 2530 (1966).

In some instances, even the introduction of new functionality can present only strategic problems. Consider a synthesis of 6-methoxy-α-tetralone (37) from benzene. The new functional group—the aromatic methoxyl—may be introduced at several stages through the sequence:

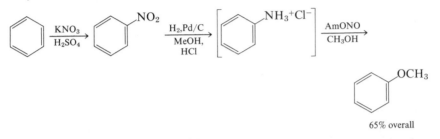

65% overall

This is a good yield process, and the only point that need be considered further is the stage at which we apply this technique. The structure of the desired end product dictates the following strategy:

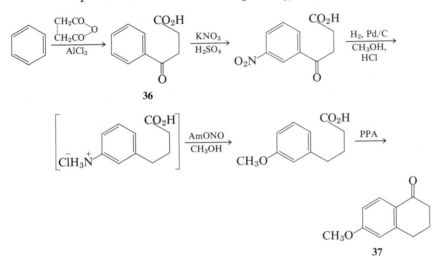

Had a different sequence of stages been chosen, the desired functional array present in 6-methoxy-α-tetralone (37) would not have resulted. If the methoxyl had been introduced in the very beginning, the anisole so formed would have been succinoylated *para* to the methoxyl grouping and

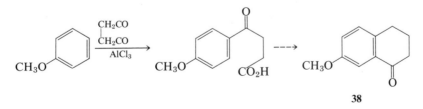

38

would have led eventually to 7-methoxy-α-tetralone (38). Similarly, had γ-phenylbutyric acid been subjected to the nitration-reduction-diazotization sequence, the predominant product expected would again be the 7-methoxy isomer 38. The desired result is possible only because β-benzoylpropionic

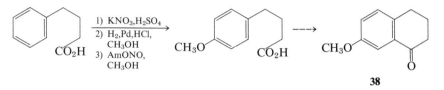

38

acid (36) nitrates *meta* to the carbonyl-containing side chain, and the directing carbonyl group can be removed during hydrogenation. The problem of introducing the new functionality *per se* is not difficult; however, to obtain the desired arrangement of groups, it is necessary to design the synthetic approach carefully.

More functionally complex problems require more attention to the scheduling of functional group addition. To this extent, the type and disposition of the functional groups required will have even more bearing on how the carbon skeleton is constructed. Thus, the more diverse and/or sensitive the functional groups become, the more the carbon-to-carbon bond-forming process must be tailored to the desired end product, and the less latitude is available in the skeletal construction phase of the synthesis. Consider, for example, the total synthesis of a molecule like penicillin V (39),[10] that "diabolical concatenations of reactive groupings" that is of such great value to man.

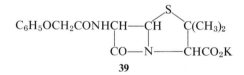

39

This molecule is an extreme case and is, in reality, nothing but diverse and sensitive functionality. The most sensitive function in the molecule is the β-lactam portion; but the remainder of the molecule is not inert to chemical change, as was the kaurene (35) skeleton above. The highly strained character of the four-membered ring of the β-lactam ring system converts the usually unreactive amide carbonyl into a highly susceptible site for attack by nucleophiles. This portion of the molecule is essentially stable only under neutral conditions and is rapidly and irreversibly cleaved by mild acid or base.[10] Any synthetic plan must take into account this overall sensitivity of the molecule due to the reactive groupings present, and, in particular, to the β-lactam function. Whenever this latter grouping is

[10] J. C. Sheehan and K. R. Henery-Logan, *J. Am. Chem. Soc.,* **81,** 3089 (1959).

formed, the conditions employed in subsequent stages are severely restricted because they must be mild enough to preserve the β-lactam system.

One approach to the solution of this problem is to grade the types of functionalities present in terms of their sensitivities and reactivities toward anticipated reaction conditions. In this manner, the carbon-to-carbon bond-forming reactions can be chosen to be compatible with the functionality as it is introduced. The least sensitive groupings are incorporated first when reaction conditions are the most vigorous. As the basic skeleton takes shape and less vigorous conditions are necessary, the more labile groupings may be included. Finally, when a very labile grouping is involved, the scheme may be tailored so as to construct that portion at the very last. Such was the plan evolved by Sheehan[10] for the synthesis of penicillin V (**39**) shown below.

Stage

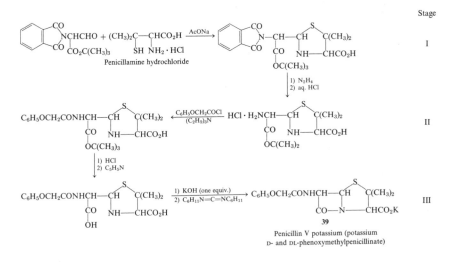

formed, Penicillamine hydrochloride

Penicillin V potassium (potassium
D- and DL-phenoxymethylpenicillinate)

The thiazolidine ring system—itself sensitive to certain hydrolytic conditions, but not nearly as labile as the β-lactam structure—is formed in stage I. The appropriate acyl residue—here, phenoxyacetyl—is then attached to the primary amine in stage II, and finally, in the last stage (III), the very labile β-lactam is closed. All of these reactions are carried out under very mild conditions and do not involve the formation of one carbon-to-carbon bond. All the required carbon-to-carbon bonds are already formed in the starting materials where functionality is minimal and stable. The remainder of the synthesis accomplishes the tricky process of uniting the labile functional groups.

The penicillin molecule is, then, the other extreme situation found in synthesis where the character of the desired functionality governs the synthetic design. Between the labile penicillin system **39** and the functionally stable kaurene (**35**) and 6-methoxy-α-tetralone (**37**) arrangements lie most of the organic molecules of synthetic interest. To a greater or lesser degree, their functionality will have a bearing on their skeletal synthesis.

EXERCISE 10

Suggest an appropriate order for the introduction of the functional groups in the following molecules.

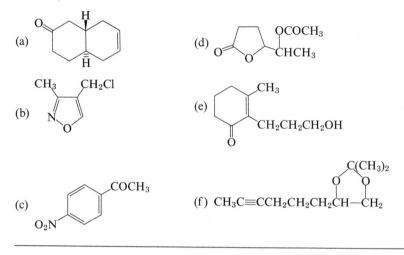

We have just discussed one method by which a synthetic plan may take ultimate functionality into account—namely, by grading the reactive groupings so as to introduce the most labile one last. Another approach to this problem of the control of reactive sites during a synthesis is to mask the grouping until it will survive the proposed further stages or until the synthesis is complete. The basic concept here is to introduce the functional group at a convenient stage and then to block that group toward further transformations with a sufficiently inert derivative. This use of *blocking groups* is a powerful and widespread concept in synthesis, for it is through their use that synthetic selectivity is made possible where none existed before.

Two examples of functional group blocking are part of the Sheehan penicillin V (**39**) synthesis. In its free form, the primary amino group in the *t*-butyl phthalimidomalonaldehydate starting material is not compatible with the aldehydo grouping. Were these two reactive functions free, polymerization of this reactant would certainly take precedence over the desired condensation with penicillamine hydrochloride. To prevent this transmogrification, the amino grouping is blocked by conversion to the phthalimide shown. The desired reaction with penicillamine hydrochloride is then observed, and the thiazolidine ring system results. Any useful blocking group must not only be stable to the proposed reaction conditions but also readily removable at the desired stage. Such is the case here, for treatment of the thiazolidine derivative with hydrazine easily frees the primary amine function after the aldehydo group has been consumed. At this stage, the amine is stable toward the other functional groups present and is now available for acylation with the desired phenoxyacetyl group.

The other blocking group employed in this synthesis is the *t*-butyl ester. This ester, present in the original *t*-butyl phthalimidomalonaldehydate and retained until the penultimate stage, serves to prevent this carboxyl group from complicating intermediate stages through reactions at this site. Any ester would accomplish this task; however, in the later stages of the penicillin synthesis, the molecule is much too labile to permit normal ester hydrolysis in order to obtain the free carboxylic acid. Therefore, rather than using a methyl or ethyl ester, Sheehan chose the base stable, acid labile *t*-butyl ester to block this carboxyl group. Mild and brief acid treat-

$$R-\overset{\overset{\displaystyle O}{\|}}{C}-O-C(CH_3)_3 \underset{}{\overset{+H^+}{\rightleftharpoons}} R-\overset{\overset{\displaystyle OH^{\oplus}}{\|}}{C}-O-C(CH_3)_3 \longrightarrow R\overset{\overset{\displaystyle OH}{|}}{C}=O + {}^{\oplus}C(CH_3)_3$$

$${}^{\oplus}C(CH_3)_3 \overset{-H^+}{\longrightarrow} CH_2=\overset{\overset{\displaystyle CH_3}{|}}{C}-CH_3$$

ment of a *t*-butyl ester results in alkyl-oxygen cleavage and generation of the *t*-butyl carbonium ion. Further fragmentation of this cation forms neutral, unreactive by-products, such as isobutylene, and renders the cleavage irreversible. Thus the *t*-butyl ester serves as an efficient carboxylic acid blocking group that is stable toward base (acyl-oxygen fission) but labile to acid (alkyl-oxygen fission).

There are as many blocking groups as there are functional groups, and there are many more simple functional group transformations which, during a given synthesis, render a particular reactive site inert. Indeed, in some instances, the proper choice of functional group masking will make the difference between success and failure in a synthesis. It is not proper in the present context to discuss at length all of the various methods used to block functionality during a synthesis.[11] Another example will suffice to emphasize the concept, and throughout subsequent discussions, the use of this approach to synthesis will be underscored.

Consider the synthesis of the hydroxyketone **43** from the commercially available diketone **40**. The problem presented by this transformation is one of designing a scheme that will allow us to modify the α,β-unsaturated ketone portion of the molecule without changing the saturated ketone

[11] J. F. W. McOmie, "Protective Groups" in *Advances in Organic Chemistry: Methods and Results* (New York, N.Y.: Wiley-Interscience Publishers, Inc., **3**, 191, 1963); *Steroid Reactions*, C. Djerassi, Ed., Holden-Day, Inc., San Francisco, Calif., 1963.

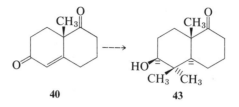

40   43

function. Inasmuch as we wish to retain the saturated ketone throughout, the best solution is to mask this portion of the molecule at the outset and retain the blocking group until the necessary differentiation between the two similar functional groups has been accomplished. To choose the appropriate ketone blocking group, we must first propose a set of reactions which will effect the desired changes at the α,β-unsaturated ketone site. One such set of reactions[12] is schematically presented as:

$$O=\overset{|}{C}-CH=\overset{|}{C}-CH_2-\quad\xrightarrow[t\text{-BuOH}]{\underset{CH_3I}{KOtBu}}\quad O=\overset{|}{C}-\underset{CH_3}{\overset{|}{C}}-\underset{CH_3}{\overset{|}{C}}=CH-\quad\xrightarrow[THF]{LiAl(OtBu)_3H}$$

$$HO-\overset{|}{CH}-\underset{CH_3}{\overset{|}{C}}-\underset{CH_3}{\overset{|}{C}}=CH-\quad\xrightarrow{H_2,Pd/C}\quad HO-\overset{|}{CH}-\underset{CH_3}{\overset{|}{C}}-\underset{CH_3}{\overset{|}{CH}}-CH_2-$$

The first thing we notice about this scheme is that all the reaction conditions are either neutral or basic—no acidic reagents are used. Therefore, if we are to mask the saturated ketone during this transformation, we must choose a blocking group that is stable toward base and labile toward acid. An appropriate such grouping is the ketal.

As appropriate as the ketal is for masking the saturated ketone during this set of basic reactions, we must still ascertain whether it can be introduced satisfactorily. In the case at hand, there is very good reason to believe that this will be possible, for while we are dealing with a diketone, one is saturated and the other is unsaturated. The effect of the double bond is to neutralize the positive character of the carbonyl carbon of the adjacent keto group by delocalization through the π-electron cloud. Thus, the α,β-unsaturated ketone is less reactive toward nucleophilic attack than the saturated ketone, and acid catalyzed ketalization with ethylene glycol should occur at a faster rate at the saturated carbonyl. If this rate difference is great enough, we will be able to effect the desired monoketalization. In fact, there is a sufficient difference in rate, and the hydroxyketone 43 can be prepared[13] as follows:

[12] See, for example, R. B. Woodward, A. A. Patchett, D. H. R. Barton, D. A. J. Ives, and R. B. Kelly, *J. Chem. Soc.*, 1131 (1957).

[13] R. E. Ireland, unpublished results.

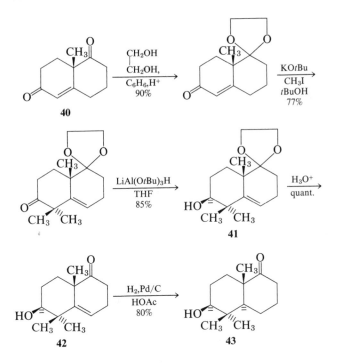

Notice that, after the hydride reduction step, the ketal has served its purpose. And although the double bond still remains, the carbonyl groups have been differentiated in the desired sense. Therefore, there is no reason to retain the ketal beyond this stage, as it can only hinder the saturation of the double bond and will provide no further useful service. Removal of a blocking group at the earliest convenient stage is generally the best policy. As we shall see in a moment, were the ketal retained during hydrogenation of the double bond, we might have predicted a different stereochemical outcome.

EXERCISE 11

Suggest a suitable blocking group that is compatible with the indicated transformation of the following molecules.

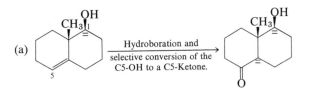

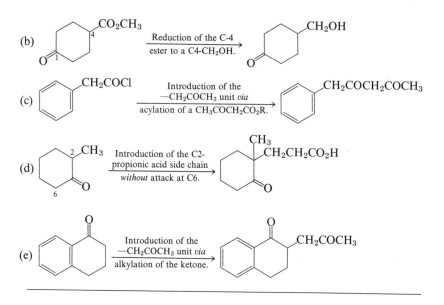

(b) Reduction of the C-4 ester to a C4-CH_2OH.

(c) Introduction of the $-CH_2COCH_3$ unit *via* acylation of a $CH_3COCH_2CO_2R$.

(d) Introduction of the C2-propionic acid side chain *without* attack at C6.

(e) Introduction of the $-CH_2COCH_3$ unit *via* alkylation of the ketone.

These examples demonstrate two important methods of handling functionality during synthesis—the introduction of labile functions last and the use of suitable blocking groups. Use of these concepts in conjunction with the requirements of carbon skeletal construction will facilitate the design of a synthetic scheme.

D. STEREOCHEMICAL CONSIDERATIONS

Another important characteristic of any molecule that must be considered in synthesis is the three-dimensional aspect of the structure. Any synthetic plan must take into account such problems as steric crowding, and geometrical and optical isomerism. All of these are the result of the spatial situation represented by the desired structure. To evaluate fully the problems presented by the spatial requirements of a molecule, we must consider the structure in three dimensions. This is done best through the use of molecular models, and many subtle structural features will be revealed in this manner. However, it is also important to represent as faithfully as possible as many of the spatial features of the system on paper as one can, for communication through molecular models is inconvenient. In any synthetic scheme which involves serious steric or stereochemical problems, the conclusions drawn from attempts to represent the three-dimensional molecule on the two dimensions of paper should be checked with models.

A classical example of the effect of steric crowding on the outcome of a chemical reaction is the case of acetymesitylene (**44**). The two *ortho*-methyl groups in this compound so crowd the acetyl grouping that the

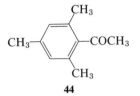

44

carbonyl cannot lie in the plane of the benzene ring. The ultraviolet spectrum of acetylmesitylene (λ_{max}^{EtOH} 242 mμ, ϵ 3600) reflects this crowding, for the nonplanarity of the aromatic ring and the carbonyl group prevents the maximum electron delocalization possible and results in the observed reduction of the extinction coefficient relative to acetophenone (λ_{max}^{EtOH} 240 mμ, ϵ 13,000). The flanking methyl groups also prevent attack of nucleophiles on the carbonyl carbon of acetylmesitylene (44), and most Grignard reagents, for instance, fail to add in the normal fashion.

EXERCISE 12

Indicate the initial product formed after the addition of methylmagnesium bromide to acetylmesitylene (44). Provide a rationalization for why the reaction between allylmagnesium bromide and acetylmesitylene results in normal 1,2-addition to the carbonyl group.

Another example of the effect of steric hindrance is provided by the hydroxyketal intermediate 41 in the synthesis discussed above. At this stage of that synthesis, two further transformations remained—namely,

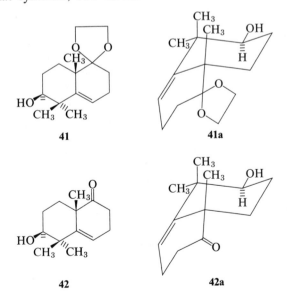

41

41a

42

42a

deketalization and hydrogenation. We chose to do them in that order on other than steric grounds, and we could have reinforced this decision had we considered the steric situation.

Catalytic hydrogenation is quite susceptible to steric hindrance, for the substrate must be adsorbed on the surface of the catalyst and any congestion on the adsorbing face of the molecule will hinder this process. Examination of the three-dimensional depiction **41a** of the hydroxyketal **41** (or better, the molecular models) reveals that both sides of the molecule are cluttered with projecting substituents. The top, or β, face of the molecule bears two axial methyl groups, and the bottom, or α, face has one of the ketal oxygen atoms oriented in an axial fashion. These axial (or perpendicular) substituents hinder close approach to either side of the hydroxyketal **41,** while the equatorial (or lateral) substituents (one methyl group and two oxygen atoms) lie in the plane of the ring system and offer little or no hindrance. The conclusion drawn from this analysis is that the hydroxyketal **41** will have difficulty approaching the hydrogenation catalyst on both sides, and the reduction will proceed slowly at best and probably randomly on both sides of the molecule. Since the desired end product of this synthetic scheme is the *trans*-fused ring system with the fusion hydrogen introduced on the α-face, *trans* to the angular methyl group, a slow, nonstereoselective hydrogenation cannot be tolerated.

Now examine the three-dimensional representation **42a** of the hydroxyketone **42.** The β-face of this molecule is still hindered by the same two axial methyl groups, but the axial oxygen atom on the α-face has been removed. There are now no protruding groupings on the α-face of the molecule, and adsorption on this side should not be hindered. The β-face of the molecule remains highly hindered by the two axial methyl groups, and consequently, we can expect little or no hydrogenation on this side. The result of this steric analysis is that hydrogenation of the hydroxyketone **42** will take place rapidly and almost exclusively on the α-side of the system. This is, indeed, the desired outcome, for the product will be the hydroxyketone **43.** Here we see the important effect of steric factors on the design of a synthetic scheme. Without taking these consequences of the three-dimensional structure into consideration at the planning stage, accurate prognosis of the experimental outcome is not possible. The steric requirements of most organic reactions can be evaluated at least qualitatively. This evaluation, taken in conjunction with the stereostructure of the proposed substrate, will generally allow one to come to a rational conclusion as to the outcome of the process on steric grounds.

In addition to juggling steps in a synthesis in order to control steric factors, as just described, one can overcome steric hindrance by arranging to form certain bonds by intramolecular reaction sequences rather than intermolecular ones. The steric congestion about a reactive site is not nearly as prohibitive for an intramolecular process as it is for the intermolecular counterpart. As a case in point, consider the preparation of the

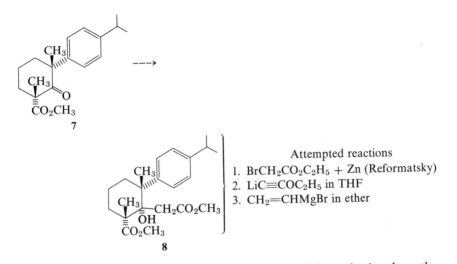

7

Attempted reactions
1. $BrCH_2CO_2C_2H_5$ + Zn (Reformatsky)
2. $LiC{\equiv}COC_2H_5$ in THF
3. $CH_2{=}CHMgBr$ in ether

8

diester **8,** an intermediate in a dehydroabietic acid synthesis where the diacid **47** is the key substance. A readily available precursor for this compound is the ketoester **7** to which it is only necessary to add the acetic acid side chain. However, even cursory inspection of this ketoester **7** reveals that the carbonyl group is highly hindered by the four adjacent substituents. It is not surprising to find that several typical sources of the required two-carbon side chain did not effect the desired change. The ketone carbonyl is just too hindered to allow the close approach of these nucleophilic reagents necessary to effect reaction.

When the system is changed to the diketone **45,** attack of the cyclic ketone can now be affected by an intramolecular reaction. Formation of the side chain enolate anion with potassium *t*-butoxide causes aldol-type

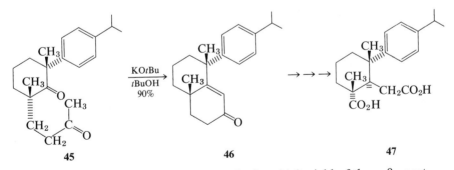

45   46   47

condensation and dehydration and results in a high yield of the α,β-unsaturated ketone **46.** This process establishes the carbon-to-carbon bonds necessary for the construction of the diacid **47,** and several further stages accomplish this transformation.[14] Thus, by a change in the synthetic

[14] R. E. Ireland and R. C. Kierstead, *J. Org. Chem.,* **31,** 2543 (1966).

design so as to allow the use of an intramolecular 1,2-carbonyl addition process, we are able to construct the desired intermediate that was unattainable when intermolecular 1,2-carbonyl addition reactions were employed. The ring carbonyl of the diketone **45** is still flanked by four substituents. But in this second approach, the key carbon-to-carbon bond-forming process occurs within the same molecule, in contrast to the bringing together of two separate reactants as was necessary before. Such an approach to the circumvention of severe steric problems may prove quite useful in synthetic planning.

EXERCISE 13

Draw the stereoformulas of the following compounds, and predict the stereochemical outcome of the indicated reactions.

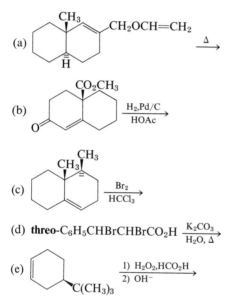

(a)

(b) $\xrightarrow{\text{H}_2,\text{Pd/C}}{\text{HOAc}}$

(c) $\xrightarrow{\text{Br}_2}{\text{HCCl}_3}$

(d) **threo-**$C_6H_5CHBrCHBrCO_2H$ $\xrightarrow{\text{K}_2\text{CO}_3}{\text{H}_2\text{O}, \Delta}$

(e) $\xrightarrow{\text{1) H}_2\text{O}_2,\text{HCO}_2\text{H}}{\text{2) OH}^-}$

STEREOSELECTIVITY

Another feature of the spatial requirements of organic molecules is, in many instances, less readily influenced. The stereochemical control of a synthetic scheme results from less well-defined methods and depends in the most part on the reactions chosen and the substrates to which they are applied. The recognition of the importance of stereoisomers in completely defining natural product structures has led to greater concern about the stereochemical outcome of organic reactions. The modern synthetic scheme—whether it is designed to lead to a natural substance or one of purely theoretical interest—must play close attention to the generation of the desired stereochemical features. Ideally, the best synthesis is a completely stereorational as well as stereoselective approach. The arithmetic demon dictates stereoselectivity for long synthesis because the loss of any portion of the intermediate compounds through the production of stereoisomeric mixtures cannot be tolerated on logistic grounds. Furthermore, sufficient attention paid to the stereochemical features of a given structure at the design stage of any synthesis may well save long, tedious laboratory time during execution of the scheme. Unfortunately, there is no pat answer for the design of a stereorational-stereoselective synthesis. In most instances, the successful approach will depend on the solution to problems unique to the products desired. We can record some generalizations here and describe some solutions to specific stereochemical problems. But the best answer to the control of stereochemistry is to pay close attention to the problem and to know well the intimacies of the organic reactions available. The stereochemical scope and limitations of many organic reactions are continually being developed, and close scrutiny of these results is worthwhile. As the synthetic objectives become more demanding, more attention is paid to stereochemical control of reactions and more is, in turn, learned about how to attain a desired result.

The control of the geometry of a double bond in an acyclic system can be a difficult process. Part of the problem is related to the fact that acyclic systems have no fixed conformation of the carbon chain. While there may be a preferred conformation for a particular reaction at the transition state, this is usually not the exclusive one, and a mixture of geometrical isomers results—as well as positional isomers, in some instances. Dehydration and dehydrohalogenation reactions are examples of reactions wherein such problems lie. A powerful method for the synthesis of olefins is the Wittig reaction, but even this method generally results in geometrical isomers in simple acyclic situations. While these isomers can be (and are) separated in many cases, a more efficient procedure would be to generate the olefins stereoselectively.

Consider the synthesis of *trans*-5,9-decadienol (**49**).[15] The *trans*-disubstituted ethylene is desired, and neither a dehydrohalogenation nor a Wittig condensation can be expected to generate solely this isomer. For the synthesis of such 1,2-disubstituted ethylene derivatives, a very successful solution to the synthetic problem is to employ an acetylenic derivative as an intermediate. Acetylene may be conveniently alkylated successively to replace both active hydrogen atoms; thus, it serves as a very useful starting material. However, more important for the present discussion, acetylenic derivatives may be reduced stereoselectively to either *cis*- or

[15] W. S. Johnson, D. M. Bailey, R. Owyang, R. A. Bell, B. Jacques, and J. K. Crandall, *J. Am. Chem. Soc.*, **86**, 1959 (1964).

trans-disubstituted ethylenes. In the case in hand, reduction of the acetylenic derivative **48** with lithium in liquid ammonia affords an excellent yield[15] of the desired *trans*-olefin **49**. Correspondingly, reduction of the same derivative **48** with hydrogen over a poisoned palladium catalyst (the Lindlar catalyst) results in the stereoselective production of the *cis*-isomer

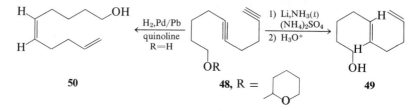

50.[16] These reduction conditions are generally applicable for the production of specific ethylenic isomers; however, notice that the method is limited to *disubstituted ethylenes which bear only groups stable to the necessary reducing media.* These may be severe limitations, but when the sequences are applicable, they represent very effective procedures.

Another method for acyclic geometrical isomer control that is useful for more highly substituted olefins also is to employ cyclic intermediates. Due to the required bond angles of the olefinic linkage, only a *cis* double bond can be accommodated in a six-membered ring system. Numerous reaction sequences can be employed to generate the familiar six-membered ring, but one of the most powerful is the Diels–Alder reaction. The Diels–Alder sequence may be used to prepare compounds which have both a double bond and suitable groupings to provide for eventual ring cleavage. Therefore, the philosophy of this approach to acyclic isomer control is to construct the cyclic olefin that most closely resembles the substitution pattern of the desired acyclic product, and then cleave the ring at an appropriate position so as to retain the *cis*-geometry of the olefin. This method has been quite successfully used by Eschenmoser to prepare the *cis*-diacid (**52**)[17] from isoprene and tetraethyl ethylenetetracarboxylate.

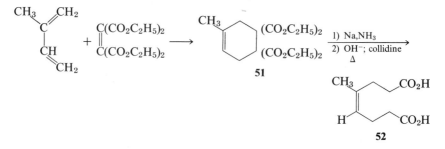

[16] W. S. Johnson and J. K. Crandall, *J. Am. Chem. Soc.,* **86,** 2085 (1964).

[17] E. Bertele, H. Boos, J. D. Dunitz, F. Elsinger, A. Eschenmoser, I. Felner, H. P. Gribi, H. Gschwend, E. F. Meyer, M. Pesaro, and R. Scheffold, *Angew. Chem.,* **76,** 393 (1964).

The intermediate cyclohexyl derivative (51) was cleaved, in this case, by an interesting reductive procedure, so as to generate initially the acyclic tetraester.

EXERCISE 14

Explain the mechanism of the cleavage step employed to open the tetraester **51**.

This approach provides entry into the trisubstituted olefins which are not available by reduction of acetylenic intermediates. The same cyclic method would be applicable to di- and tetrasubstituted ethylenes as well. Again there are limitations. For example, since this sequence employs cyclic intermediates, only *cis*-olefins are available.

The two foregoing examples are representatives of two valuable methods for the control of geometrical isomerism in synthesis. They are by no means the only approaches to the problem; they only illustrate two modes of attack. By their example, we should recognize that in approaching the synthesis of olefinic molecules—particularly acyclic ones—geometrical isomerism and its control are part of our concern.

Finally, we must face the problem of optical isomerism. Herein lie two porisms. First, we may look at the resolution of synthetic compounds into their enantiomeric forms. Unless optically active intermediates are used, the construction of organic molecules which contain asymmetric carbon atoms will generate a racemic end product. The final completion of such a synthesis, then, entails the resolution of this racemic mixture. There is little theory associated with this process, and whenever such a resolution is contemplated, it is best to rely on the formation of diastereoisomeric derivatives with an optically active reagent. For instance, $(\pm)$-penicillin V, (39) discussed earlier was resolved through the agency of the enantiomers of **erythro**-1,2-diphenyl-2-methylaminoethanol (53).[10] This is an example of the use of salt formation between racemic acid and optically active base. The advantage of this approach over that where covalent bonds are formed

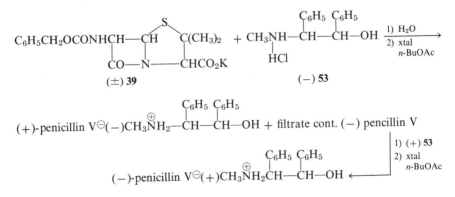

is the ease of diastereoisomeric derivative formation and cleavage. However, not all salts will result in resolution of a racemic mixture, and another method (bioresolution, derivativization by covalent bond formation, etc.) must be tried. In many cases, the history of the racemic molecule will suggest the best choice. Here, the natural penicillins—specifically, penicillin G—had been employed to effect the resolution of the racemic amine **53**, and therefore, the optically active amine **53** was a natural choice to resolve the racemic penicillin V (**39**). When such experience is not available, the resolution process is one of trial and error, in which several attempts may be necessary.

A considerably more important problem from the synthetic standpoint is the control of asymmetry during the construction of polyasymmetric molecules. *Recognize that any synthetic sequence which begins with non-asymmetric or racemic starting materials and employs racemic reagents will ultimately generate only racemic products.* However, when a new asymmetric center is introduced into a molecule which is itself asymmetric, two racemic diastereoisomeric products can theoretically result. The goal of a stereoselective synthesis is to arrange the synthetic scheme so that one of the racemic diastereoisomeric pairs predominates, preferably to the exclusion of the other. Consider the case of the synthesis of the hydroxyketone **43** that we discussed earlier. Our synthesis began with a racemic starting material that contained one asymmetric center (C9); but through the scheme we introduced two new asymmetric centers—the carbons C-6

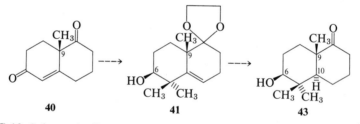

40   41   43

and C-10. Schematically, we can represent the theoretically possible stereoisomeric results as follows (the superscripts refer to the particular asymmetric carbon atom involved):

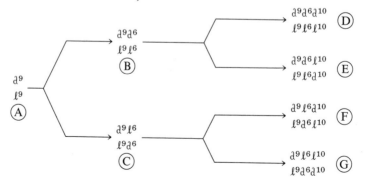

The result is that the synthesis could generate eight enantiomers and, more pertinent, the four racemic diastereoisomeric pairs ⒟, ⒠, ⒡, and ⒢. Obviously, such a result cannot be tolerated from either a practical or aesthetic standpoint. A method for stereochemical control must be recognized at each stage that generates a new asymmetric center.

We may first simplify the problem by realizing that we are dealing with racemic materials. Thus racemate Ⓐ—the unsaturated diketone **40**—can produce racemates Ⓑ and Ⓒ—the hydroxyketal **41**—which in turn can lead to racemates ⒟, ⒠, ⒡, and ⒢—the ketalcohol **43**. All are diastereoisomeric and therefore possess different physical and chemical properties. Consider the first transformation: Ⓐ to Ⓑ and/or Ⓒ. This involves the reduction of the C-6 carbonyl, and we can represent this change chemically as follows:

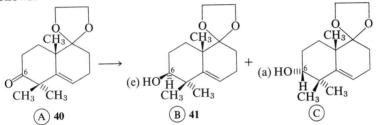

Notice that, in each case, only one of the two possible enantiomers is drawn. We have specified earlier that we are dealing with racemic compounds and in order to present graphically the particular change that has occurred, and not confuse the presentation with too many formulas, we draw only one of the enantiomers. *We must recognize that, in spite of the specific drawing, Ⓐ, Ⓑ, and Ⓒ occur as racemic compounds due to our original stipulation, and they are accompanied by their antipodes.* We will use this convention throughout.

Our problem now is to devise some method whereby more—and preferably, only—Ⓑ is formed on reduction of Ⓐ. Such stereoselectivity can be accomplished if we recognize that the C-6 ketone is relatively unhindered and that reduction with a chemical reagent like lithium aluminum tri-*t*-butoxy hydride will result in product development control, and thereby, the more stable isomer will be formed. An equatorially oriented substituent on a cyclohexane ring is considerably more stable than its axially oriented counterpart; hence, we can expect the racemate Ⓑ—equatorial hydroxyl—to predominate over racemate Ⓒ—axial hydroxyl. Were we to choose another method of reduction such as catalytic hydrogenation, this might not be the expected result, and a different ratio of isomers Ⓑ/Ⓒ might be formed. Catalytic hydrogenation of such a ketone will usually produce the more sterically congested (axial) hydroxyl function, since adsorption of the carbonyl function on the catalyst will preferentially occur on the less hindered side, and delivery of hydrogen to that side will force the resulting hydroxyl group onto the opposite more

4

Wherein the Carbon Skeleton Is the Thing

2,4-DIMETHYL-2-HYDROXYPENTANOIC ACID (54)

While it will become obvious that certain intermediates in this synthesis are commercially available, for practice let's consider how this acid may be prepared from one-carbon fragments and any desired inorganic materials and solvents. With these ground rules, we tackle the synthetic design by working back, step by step, to starting materials from our desired objective.

$$
\underset{\textbf{54}}{\overset{\displaystyle CH_3 \quad OH}{CH_3CHCH_2\underset{\underset{\displaystyle CH_3}{|}}{C}\!\!-\!\!CO_2H}} \xleftarrow[\underset{60-70\%}{2)\ conc.\ HCl,\ \Delta}]{1)\ NaHSO_3;\ NaCN} \underset{\textbf{55}}{\overset{\displaystyle CH_3 \quad O}{CH_3CHCH_2CCH_3}}
$$

We must first recognize that a carbon chain gains its chemical reactivity toward most reagents through the presence of a functional group. Therefore, to devise a method for the construction of the desired carbon skeleton, it is best to concentrate first on the functionality present. We should first reduce the size of the carbon chain so that it may be reconstructed as easily as possible. The α-hydroxyacid arrangement in the present objective suggests that this grouping might best arise from hydrolysis of a cyanohydrin. The latter substitution is most readily obtained through the addition of hydrogen cyanide to a ketone derivative.[1] If we apply this conclusion to the acid **54**, a suitable precursor would then be the ketone **55**. Treatment of methyl isobutyl ketone (**55**) in an aqueous medium with sodium bisulfite and sodium cyanide will generate the cyanohydrin which, without purification, may be hydrolyzed in hot concentrated hydrochloric acid to the desired acid **54**.

Our attention now shifts to the construction of the ketone **55**. Here again we can best plan the synthesis of this material by working around the carbonyl function.

---

[1] For a representative procedure, see W. G. Young, R. Dillon, and H. J. Lucas, *J. Am. Chem. Soc.*, **51**, 2528 (1929).

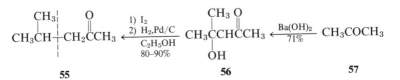

$$\underset{\textbf{55}}{CH_3CH\text{---}CH_2CCH_3} \xleftarrow[\substack{C_2H_5OH \\ 80\text{–}90\%}]{\substack{1)\ I_2 \\ 2)\ H_2,Pd/C}} \underset{\textbf{56}}{CH_3CCHCCH_3} \xleftarrow[71\%]{Ba(OH)_2} \underset{\textbf{57}}{CH_3COCH_3}$$

Consideration of the structure of this ketone will reveal several methods by which it might be made. Carbons might be added, one by one, from isobutyl bromide, for instance, but a more efficient method is revealed if we consider the entire carbon skeleton. The molecule may be divided into two three-carbon fragments, each bearing functionality on the central carbon atom. The functionality of one three-carbon unit must be used to join the two fragments together, and that of the other unit must be retained as the ketone. If both fragments contained carbonyl groups as the central carbon, we would have two identical three-carbon units as precursors, namely, acetone (**57**). This would be an ideal solution to the construction of this ketone, for it allows the construction of the desired carbon skeleton through the union of its two halves. We next need to consider how this union might be accomplished. To this end, the self-condensation of acetone[2] in the presence of base will suffice to put the carbon skeleton together. This leaves only the problem of functional group modification before the desired saturated ketone **55** may be obtained. The removal of the superfluous hydroxyl group of the condensation product **56** (diacetone alcohol) may best be achieved through a two-stage sequence of dehydration[3] and catalytic hydrogenation of the resulting α,β-unsaturated ketone.

To solve the problem as described, we need only devise a synthesis of acetone (**57**) from one-carbon fragments. As trivial as this preparation is,

$$\underset{\textbf{57}}{CH_3COCH_3} \xleftarrow[400\text{–}500°]{thoria\ cat.} CH_3CO_2H \xleftarrow[\substack{2)\ CO_2 \\ 3)\ H_3O^+}]{1)\ Mg,\ ether} CH_3I$$

it would be important if we were interested, for instance, in a C^{14}-labelled acetone molecule. Again the approach is similar in that dissection of the acetone molecule is done at the functional group. Thus we see that acetone becomes available from acetic acid which, in turn, is available from methyl iodide by carboxylation of the Grignard reagent. Our synthetic plan is now complete and available for laboratory execution. The scheme so devised is the result of working from the complex to the simple in a step-by-step fashion and is an example of the general methodology of synthetic planning. A multistep synthesis becomes nothing more than a series of one- or two-step sequences.

[2] J. B. Conant and N. Tuttle, *Org. Syn. Coll.,* Vol. 1, 199 (1941).
[3] J. B. Conant and N. Tuttle, *ibid.,* 345 (1941).

5-HEXENOIC ACID (58)

This difunctional material is again primarily an example of carbon skeletal construction. The functional groups are quite diverse, and we can anticipate little difficulty in the choice of selective reaction conditions. Consider how the acid **58** might be made from readily available starting materials.

$$CH_2{=}CHCH_2CH_2{\mid}CH_2CO_2H \xleftarrow[\substack{3)\ 160°(-CO_2) \\ 65\%}]{\substack{1)\ NaOC_2H_5 \\ 2)\ NaOH}}$$

$$\textbf{58}$$

$$CH_2{=}CHCH_2CH_2Br + CH_2(CO_2C_2H_5)_2$$
$$\textbf{59}$$

Let us first examine the process by which the acid **58** was first made by Linstead and Rydon.[4] We approach the problem by simplifying the desired end product in a step-by-step fashion. The two functional groups are at the termini of a straight carbon chain. They may therefore be either added individually to a three-carbon chain or incorporated with one or more methylene groups and these larger fragments joined. The latter process is logistically the better approach, for it entails the coupling of larger subunits rather than the piecemeal addition of carbons. If we regard the 5-hexenoic acid (**58**) as a butenyl acetic acid, a method for its construction through the alkylation of acetic acid becomes apparent. Inasmuch as direct alkylation of acetic acid or its derivatives is fraught with experimental difficulties, the substitution of diethyl malonate is suggested.[5] The active methylene of diethyl malonate is readily alkylated, and through hydrolysis and decarboxylation, the alkylated malonate serves as an ideal precursor of the desired acetic acid derivative. Thus the role of diethyl malonate in such a synthesis is to provide a selectively reactive derivative of acetic acid, wherein the second carbethoxyl grouping facilitates the substitution process and yet may be readily removed at a later stage. In this case, the conversion of 4-bromo-1-butene (**59**) to the acid (**58**) was accomplished in a 65% overall yield by Linstead and Rydon[4] and represents a good example of the "malonic ester" synthesis of substituted acetic acids.

Since diethyl malonate is commercially available, the remainder of the synthetic planning focuses on the construction of 4-bromo-1-butene (**59**). Alkyl halides are more readily prepared from alcohols rather than as a result of direct carbon-to-carbon bond-forming reactions. Therefore, it is

[4] R. P. Linstead and H. N. Rydon, *J. Chem. Soc.*, 1994 (1934); for a more sophisticated, modern synthesis of the acid **58**, see C. J. Albisetti, N. G. Fisher, M. J. Hogsed, and R. M. Joyce, *J. Am. Chem. Soc.*, **78**, 2637 (1956). The overall yield of this latter process is not quite as high as the older procedure, but the number of transformations is fewer.

[5] C. D. Gutsche, *op. cit.*

$$CH_2=CHCH_2CH_2Br \xleftarrow[\substack{pyr. \, 0° \\ 78\%}]{PBr_3} CH_2=CHCH_2{-}CH_2OH \xleftarrow[\substack{2) \, H_3O^+ \\ 47\%}]{\substack{1) \, Mg, \, ether \\ (HCHO)_3}}$$

$$\textbf{59} \qquad\qquad\qquad\qquad \textbf{60} \qquad\qquad\qquad CH_2=CHCH_2Br$$

$$\textbf{61}$$

logical to consider the next simplification of the system to be a functional group transformation leading to allyl carbinol (**60**). In practice, this alcohol **60** was readily converted to the necessary 4-bromo-1-butene (**59**) by treatment with phosphorus tribromide in pyridine.

Contrary to alkyl halides, alcohols are commonly the result of carbon-to-carbon bond-forming reactions; thus it is appropriate at this stage to consider how the carbon skeleton of allyl carbinol (**60**) might be formed. As the name of this alcohol **60** implies, the most readily available precursors are an allyl derivative and a one-carbon fragment. Here, then, is an ideal circumstance for the application of the Grignard reaction between allyl magnesium bromide and formaldehyde. Linstead and Rydon were able to carry out this sequence in 37% overall yield by the procedures indicated earlier, and they used allyl bromide (**61**) as the commercially available starting material. Again we have been able to construct a synthetic plan by working from the complex end product to the simple, commercially available starting materials in a logical, step-by-step fashion.

Further consideration of the synthesis of the hexenoic acid **58** reveals alternate methods which might also suffice. These are particularly the result of knowledge of organic reactions gained since Linstead and Rydon first prepared the acid **58**. For instance, the intermediate allylcarbinol (**60**) might be synthesized by the interaction of vinyl lithium and ethylene

$$CH_2=CHCH_2CH_2OH \xleftarrow[H_3O^+]{then} CH_2=CHLi + \overset{O}{\overset{\displaystyle \diagup\!\!\diagdown}{CH_2{-}CH_2}}$$

$$\textbf{60}$$

oxide. This process affords the carbinol **60** through the combination of the two largest carbon fragments possible, and again both starting materials are commercially available.

One might also entertain the possibility of preparing the desired acid **58** from two more nearly equal carbon fragments. This approach might take on the character of a 1,4-(conjugate)-addition of an allyl anion **62** to a vinyl carbonyl compound **63**. Since a common source of such carbanionic species is an organometallic derivative, an allyl copper derivative formed

$$CH_2=CHCH_2{-}CH_2CH_2CO_2H \longleftarrow CH_2=CHCH_2^{\ominus} + CH_2=CH\overset{\overset{\displaystyle O}{\|}}{C}R$$

$$\textbf{58} \qquad\qquad\qquad\qquad \textbf{62} \qquad\qquad \textbf{63}$$

from the Grignard reagent and cuprous iodide may be suggested for the allyl anion **62.** The logical choice for a vinyl carbonyl compound **63** is ethyl acrylate. This ester, however, may not serve as well as the related methyl vinyl ketone **(65)** in the conjugate addition of the allyl copper reagent. Since the methyl vinyl ketone adduct **64** may just as readily be cleaved to the desired acid **58** by oxidation with sodium hypobromite as the corresponding ester could be hydrolyzed with alkali, the use of the vinyl ketone seems more advisable. This is an attractive approach to the

$$CH_2{=}CHCH_2CH_2CH_2CO_2H \xleftarrow{\text{NaOBr}} CH_2{=}CHCH_2CH_2CH_2\overset{\overset{\displaystyle O}{\|}}{C}CH_3 \xleftarrow[\substack{\text{CuI, cat.} \\ \text{2) } H_3O^{\oplus}}]{\text{1) Mg, ether}}$$
$$\textbf{58} \qquad\qquad\qquad\qquad \textbf{64}$$

$$CH_2{=}CHCH_2Br + CH_2{=}CH\overset{\overset{\displaystyle O}{\|}}{C}CH_3$$
$$\textbf{61} \qquad\qquad \textbf{65}$$

synthesis of the acid **58,** for we are again combining the two largest possible carbon fragments in one step, and the necessary starting materials are commercially available. The overall yield of 5-hexenoic acid **(58)** by the five-step synthesis used by Linstead and Rydon is 24%. While the latter modification has not been attempted, it is entirely possible that an even higher yield would result from the three-step process which directly combines the two larger fragments. Thus we see another facet of organic synthesis: there are usually several satisfactory ways to synthesize a given molecule.

δ-Caprolactone (66) [5-Hydroxyhexanoic Acid Lactone]

When we approach the synthesis of the lactone **66,** we have before us some prior art to consider. The 5-hexenoic acid **(58)** prepared earlier might suffice as starting material for the lactone construction. It is reasonable to assume that acid-catalyzed addition of the carboxyl group across the

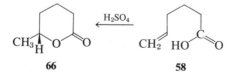

$$\textbf{66} \qquad\qquad\qquad\qquad \textbf{58}$$

terminal double bond would generate the desired δ-lactone. This six-membered ring lactone will most certainly be favored over the seven-membered ring isomer both on steric (six ring more stable than seven) as well as electrostatic (2° carbonium ion intermediate more stable than 1°) grounds. If we adopt this approach to the synthesis of the lactone **66,** the foregoing synthesis of 5-hexenoic acid **(58)** will serve to prepare the starting material, and we need only append the last sulfuric-acid-catalyzed lactonization step.

We should consider this scheme more carefully, though, particularly the acid catalyzed lactonization step. To generate the required 2° carbonium ion, we will have to employ strong mineral acid to protonate the weakly basic double bond. Under such conditions, it is reasonable to assume that the generated lactone will also protonate and hence be in equilibrium with the precursor olefinic acid. However, the olefinic acid regenerated from the lactone need not be—and probably will not be—the 5-hexenoic acid (**58**). The 4-hexenoic acid (**67**), which has a disubstituted double bond, is more stable and hence a more likely product. We have thus isomerized a terminal

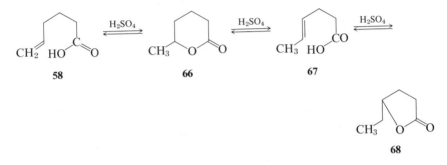

olefin to an internal position—a very reasonable result. The problem is that the 4-hexenoic acid (**67**) may now lactonize to form the five-membered ring lactone **68**, as well as undergo further acid-catalyzed isomerization of the double bond down the carbon chain toward the carboxyl group. The generation of the five-membered ring lactone (more stable than the six) and the generation of isomeric olefinic acids, as well as of polymeric material from olefin addition reactions, all augur against the obtaining of a high yield of the desired hexenyl lactone by this route. This was indeed found to be the case when Linstead and Rydon[4] carried out the experiment, for only a low yield of impure δ-lactone was formed. This result points up the care that must be given to the reaction conditions in which we generate products. In a chemical reaction, we must expect an organic molecule to do anything and everything compatible with the experimental conditions. The solution is to select milder conditions or modify the synthetic scheme so as to avoid the offending reaction.

A solution to the problem at hand is to form the lactone by the interaction of an alcohol and carboxyl group rather than an olefin and carboxyl group. Thus Linstead and Rydon[4] chose to carry out the intramolecular esterification of a hydroxyl group on C-5 rather than to cause the addition of the carboxyl group to the terminal double bond. The necessary hydroxyl group is best formed by reduction of a C-5 ketone rather than hydration of the double bond (why?); thus the synthesis of the lactone becomes the

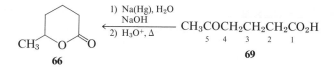

66 ← 1) Na(Hg), H_2O / NaOH / 2) H_3O^+, Δ ← $CH_3COCH_2CH_2CH_2CO_2H$
$\quad\quad\quad\quad\quad\quad\quad\quad\quad\quad\quad\quad\quad\quad\quad\quad\quad$ 5   4   3   2   1
$\quad\quad\quad\quad\quad\quad\quad\quad\quad\quad\quad\quad\quad\quad\quad\quad\quad\quad$ **69**

synthesis of the ketoacid **69**. Again we must consider how this ketoacid **69** may be prepared from readily available materials and in the fewest steps possible. As in the case of the 5-hexenoic acid (**58**), the most efficient process for such six-carbon straight-chain molecules will be the combination of two three-carbon fragments. This may be accomplished here through the alkylation of acetone with an appropriate propionic acid derivative.

$$\underset{\textbf{69}}{CH_3COCH_2CH_2CH_2CO_2H} \xleftarrow{\ H_3O^+\ } \underset{\textbf{70}}{CH_3CO\overset{\displaystyle CO_2C_2H_5}{\overset{|}{C}}HCH_2CH_2CO_2C_2H_5} \xleftarrow{\ NaOC_2H_5\ }$$

$$CH_3COCH_2CO_2C_2H_5 + ClCH_2CH_2CO_2C_2H_5$$

To obviate the experimental problems of polyalkylation of acetone itself in strong base, ethyl acetoacetate serves as the source of this ketonic fragment. The β-ketoester system allows stoichiometric enolate formation with one equivalent of sodium ethoxide and, after alkylation, the activating carbethoxyl group may be easily removed by hydrolysis and decarboxylation. While the alkylating agent chosen was ethyl β-chloropropionate, the Michael-type addition of the acetoacetate anion to ethyl acrylate would also suffice to join the two fragments. Hydrolysis and decarboxylation of the resulting keto diester **70** in aqueous mineral acid afforded Linstead and Rydon[4] an 80% yield of the desired ketoacid **69**. Inasmuch as both starting esters are commercially available, this short four-step sequence provides an efficient synthesis of the δ-lactone **66**.

BICYCLO-[4.1.0]-HEPTAN-2-ONE (71)

At the outset, this cyclopropyl ketone does not appear to be related to any of the compounds we have discussed. However, careful consideration of the structure of this material reveals that one of several possible schemes stems from 5-hexenoic acid (**58**).

A cyclopropane ring system in a molecule should bring to consideration the possibility of generating that skeletal arrangement by the addition of a carbene to a double bond. Two logical possibilities for this transfor-

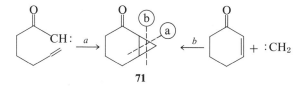

71

mation are indicated by the paths *a* and *b*. Each involves the addition of a carbene to an olefin, and while path *b* adds the three-membered ring to the six-membered ring, path *a* entails cyclization of an acyclic precursor and forms the bicyclic system **71** directly. Both schemes are suitable here, but each is distinct from the other. In more complex cases, one may be preferable to the other.

In path *a* we require an acyclic carbene precursor. A knowledge of carbene chemistry and rudimentary skeletal considerations leads us back to our old friend, 5-hexenoic acid (**58**). Using this acid **58** as a source of

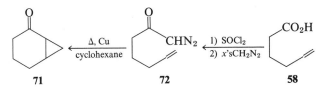

the diazoketone **72,** Stork[6] was able to show that, when the latter material was heated with copper bronze in cyclohexane solution, the bicyclic ketone **71** was formed in good yield. Such a sequence is an efficient method for the preparation of acyclic as well as cyclic cyclopropyl ketones, and the procedure has been used[7] in other more complex cases. From the standpoint of the synthesis at hand, these three stages suffice to elaborate a synthetic plan for the bicyclic ketone **71** from readily available materials, as we have already described the synthesis of 5-hexenoic acid (**58**).

Path *b* is equally suitable for the synthesis of the desired bicyclic ketone **71**. Consideration of the addition of carbene itself to an α,β-unsaturated ketone brings to mind the several methods available for the generation of divalent carbon derivatives. One of the most successful methods for the generation of such a reactive one-carbon unit is through the

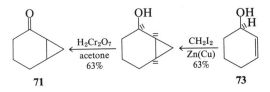

interaction of methylene iodide and a zinc-copper couple (Simmons–Smith reaction). The organometallic intermediate that results from this pair adds very efficiently to double bonds, particularly those with an adjacent hydroxyl group. By using such a sequence, Dauben[8] was able to effect the transformation of 2-cyclohexenol (**73**) to the desired bicyclic ketone **71** in 40% overall yield. Considerable work[8] has been recorded on this cyclopro-

[6] G. Stork and J. Ficini, *J. Am. Chem. Soc.,* **83,** 4678 (1961).

[7] M. M. Fawzi and C. D. Gutsche, *J. Org. Chem.,* **31,** 1390 (1966).

[8] W. G. Dauben and G. H. Berezin, *J. Am. Chem. Soc.,* **85,** 468 (1963); *ibid.,* **89,** 3449 (1967).

pylation sequence. It has been found that, not only does the allylic (or homoallylic) hydroxyl function facilitate the reaction, but it also controls the stereochemistry of the addition of the methylene group to the double bond.

2-Cyclohexenol (73) is a readily available substance, since it may be obtained readily by the reduction of commercially available 2-cyclo-

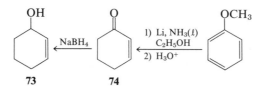

73   74

hexenone (74). However, cyclohexenone is a rather labile α,β-unsaturated ketone, and its shelf-life is not very long. For routine use of such intermediates, it is usually best to prepare them as needed, if possible. In this case, an efficient method is available for the preparation of this enone 74. The alkali metal-ammonia-alcohol reduction[9] of anisole first generates the dihydro derivative which, on acid-catalyzed hydrolysis, is converted to the desired cyclohexenone. This method for the conversion of an aromatic substance to an aliphatic material by noncatalytic, chemical means is a very powerful tool of the synthetic organic chemist. We will see it used often in more complex and delicate situations.

7-ISOPROPYL-*TRANS*-3,7-OCTADIENOL-1 (75)

This dienol represents a somewhat more challenging problem than do the foregoing molecules, as it carries more functionality and, particularly, two similar functional groups. Let us first analyze the structural and functional problems presented, and then tackle the synthetic plan.

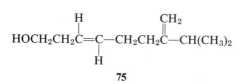

75

The features of this molecule which bear consideration are the *trans*-disubstituted ethylenic unit, the primary alcohol, and the terminal methylene. The basic skeleton is little more than a straight carbon chain; hence, the molecular construction resolves itself to one of uniting the appropriate functional groups. The order of this union is important.

---

[9] H. Smith, *Organic Reactions in Liquid Ammonia* (New York, N.Y.: Interscience Publishers, 1963).

A suitable method for the generation of a *trans*-double bond is the metal-ammonia reduction[9] of a disubstituted acetylene mentioned earlier. The terminal methylene can best be obtained through the reaction of methylenetriphenylphosphorane with the corresponding ketone. The primary alcohol can be generated by reduction of an ester group or carried along *per se* throughout the scheme in a protected fashion. The decision to introduce the *trans*-double bond by reduction of an acetylenic linkage is the key to the synthetic plan, and the remaining construction must revolve around this judgement. The fact that a *trans*-olefin in a specific location in the molecule is required makes the acetylenic approach very attractive. It is entirely reasonable to build the synthesis around this concept. The ease of substitution of acetylenic hydrogens also offers an attractive avenue for the carbon skeletal construction at early stages.

At the outset, let's not consider the source of the primary hydroxyl function until a scheme for the construction of the carbon skeleton and the diene system is outlined. At that juncture, we can evaluate the reactions chosen and decide what method is best for the inclusion of this oxygen function. Since we envisage a metal-ammonia reduction of the triple bond to generate the *trans*-double bond, the remainder of the molecule must be stable toward these reducing conditions. The reduction of the enyne **76** satisfies this requirement, for the isolated terminal double bond is not as readily reduced as the triple bond by the metal-ammonia

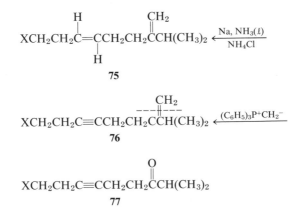

75

76

77

combination. It is a simple matter, then, to decide on the ketone **77** as the precursor for the enyne **76,** since the Wittig reaction will readily accomplish this latter transformation.

Several methods exist for the construction of the acetylenic ketone **77,** but we should concentrate on devising an approach that will unite the two largest carbon fragments. We saw earlier that ethyl acetoacetate was an ideal precursor of an acetone derivative; here, use of an analogous β-keto-ester will solve this similar synthetic problem. Alkylation of ethyl iso-

$$XCH_2CH_2C\equiv CCH_2 \!\!-\!\! CH_2\overset{O}{\overset{\|}{C}}CH(CH_3)_2 \xleftarrow[\text{2) } H_3O^+]{\text{1) NaH, DME}}$$

77

$$XCH_2CH_2C\equiv CCH_2Br + C_2H_5O_2CCH_2\overset{O}{\overset{\|}{C}}CH(CH_3)_2$$

78   **79**

$$(CH_3)_2CHCO_2H \xrightarrow[\text{N}(C_2H_5)_3]{ClCO_2C_2H_5,} [(CH_3)_2CHCO_2CO_2C_2H_5] \xrightarrow[\text{2) } NH_4Cl, H_2O]{\begin{array}{l}\text{1) } CH_2\overset{CO_2C_2H_5}{\underset{CO_2H}{<}} \text{, } Mg(OC_2H_5)_2\end{array}}$$

80

butyrylacetate (**79**) [readily available from the mixed anhydride **80** of butyric acid and the magnesium chelate of ethyl hydrogen malonate[10]] with the acetylenic bromide **78** will occur only on the activated methylene of the β-ketoester **79**. After hydrolysis and decarboxylation, the desired acetylenic ketone **77** will result.

We must now consider how the alkylating agent **78** may be prepared. Again, recognize that carbon-to-carbon bond-forming reactions do not generally result in the formation of alkyl halides. The best source of such halides is through replacement of an alcohol functional group transformation. This then suggests that the alcohol **81** is the logical precursor of the

$$XCH_2CH_2C\equiv CCH_2Br \xleftarrow[\text{pyridine}]{PBr_3} X\!\!-\!\!CH_2CH_2C\equiv C\!\!-\!\!CH_2OH \xleftarrow[\text{2) } H_3O^\oplus]{\begin{array}{l}\text{1) } C_2H_5MgBr; \\ HCHO\end{array}}$$

78   **81**

$$XCH_2CH_2C\equiv CH$$

82

desired halide **78**. At this point, we can consider a carbon-to-carbon bond-forming reaction for the preparation of the alcohol **81**. Advantage should be taken of the active hydrogen on a monosubstituted acetylene for this transformation, for metal acetylides are readily prepared, and they add efficiently to carbonyl compounds to form alcohols. The culmination of this reasoning is the choice of the butyne **82** as a suitable starting point for the synthesis. Formation of the magnesium bromide salt of the acetylenic derivative **81** by exchange with ethyl magnesium bromide, and addition of formaldehyde to this solution, will generate the alcohol **81**. Treatment of this alcohol **81** with phosphorous tribromide in pyridine will result in the formation of the desired acetylenic bromide **78**.

[10] R. E. Ireland and J. A. Marshall, *J. Am. Chem. Soc.*, **81**, 2907 (1959).

Several substituted butynes are commercially available, and we must finish the synthetic plan by deciding which of them to use. This decision need not be arbitrary. At this juncture, we have completed a scheme for the synthesis of the carbon skeleton of the dienol **75** and are ready to decide what functional group will best serve as a source of the requisite primary alcohol. At this stage, consideration of the functionality and reaction conditions necessary for the skeletal construction will limit the possible choices for the primary alcohol precursor. A carbonyl-containing functional group would be unsuitable because the carbonyl would interfere with the initial Grignard reaction with formaldehyde. Furthermore, the methylene group of this acetylenic starting material is now flanked by two

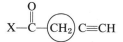

activating groups, and the resulting acidity of these hydrogens will complicate the β-ketoester alkylation stage and the Wittig reaction. A primary alcohol group still bears an active hydrogen, and as such, 3-butyn-1-ol (**83**)

$$HOCH_2CH_2C{\equiv}CH$$
83

itself will not be a suitable starting material, for it will again complicate subsequent reactions. Except for this active hydrogen, the alcohol function is an ideal substituent, since the carbon-oxygen bond will be inert to the planned reaction scheme, and ultimately, we want an alcohol at that position.

A solution would be to use an ether function in place of this alcohol, as both carbon-oxygen bonds will be inert. We must, however, provide for the subsequent cleavage of the ether and regeneration of the alcohol under mild enough conditions so that the olefinic linkages are not disturbed. An ether grouping which satisfies these requirements is an alkyl benzyl ether. The benzyl grouping may be readily removed by hydrogenolysis with hydrogen over a palladium catalyst or sodium in liquid ammonia. The latter conditions are ideal for the present case, since we plan to use these reducing conditions as a last step anyway. The choice of starting material is, then, the benzyl ether **84** which is available from 3-butyn-1-ol (**83**) and

$$C_6H_5CH_2OCH_2CH_2C{\equiv}CH \xleftarrow[\text{acetone}]{\text{K}_2\text{CO}_3} C_6H_5CH_2Cl + HOCH_2CH_2C{\equiv}CH$$
84   **83**

benzyl chloride. The resulting synthetic plan, prepared by reasoning backward, has been successfully used by W. S. Johnson and co-workers[11] and is completely outlined below.

[11] W. S. Johnson, N. P. Jensen, and J. Hooz, *J. Am. Chem. Soc.*, **88**, 3859 (1966).

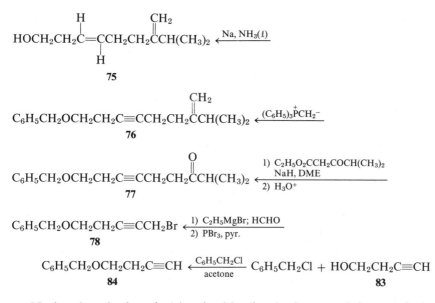

$$\underset{75}{HOCH_2CH_2\overset{\overset{\displaystyle H}{|}}{C}=\overset{\overset{\displaystyle CH_2}{\|}}{\underset{|}{C}}CH_2CH_2\overset{}{C}CH(CH_3)_2} \xleftarrow{\text{Na, NH}_3(l)}$$

$$\underset{76}{C_6H_5CH_2OCH_2CH_2C\equiv\overset{\overset{\displaystyle CH_2}{\|}}{C}CH_2CH_2\overset{}{C}CH(CH_3)_2} \xleftarrow{(C_6H_5)_3\overset{+}{P}CH_2^-}$$

$$\underset{77}{C_6H_5CH_2OCH_2CH_2C\equiv CCH_2CH_2\overset{\overset{\displaystyle O}{\|}}{C}CH(CH_3)_2} \xleftarrow[\text{2) H}_3O^+]{\substack{\text{1) C}_2H_5O_2CCH_2COCH(CH_3)_2 \\ \text{NaH, DME}}}$$

$$\underset{78}{C_6H_5CH_2OCH_2CH_2C\equiv CCH_2Br} \xleftarrow[\text{2) PBr}_3,\text{ pyr.}]{\text{1) C}_2H_5MgBr;\text{ HCHO}}$$

$$\underset{84}{C_6H_5CH_2OCH_2CH_2C\equiv CH} \xleftarrow[\text{acetone}]{C_6H_5CH_2Cl} \underset{83}{C_6H_5CH_2Cl + HOCH_2CH_2C\equiv CH}$$

Notice that the key decision in this plan is the use of the acetylenic group and that this decision was made after consideration of the entire molecule. The subsequent skeletal construction, as well as functional group manipulation, was planned around the use of this acetylenic group to overcome what appears to be the synthetically most difficult structural feature.

4,6-DIMETHOXYPHTHALALDEHYDIC ACID (85)

Most simple mono- and disubstituted aromatic compounds are readily available or may be prepared by standard procedures. Some procedures may be lengthy in order to overcome the directing effects of groups already present, but they usually entail only functional group transformations. For instance, *meta*-nitrotoluene is not available by direct nitration of toluene because the methyl group has an *ortho-para* directing effect. However, this

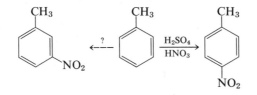

transformation may be accomplished if we take advantage of the greater directive effect of the acetamido grouping. By placing this substituent in the position *para* to the methyl group, we not only force nitration to take place *meta* to the methyl (*ortho* to the acetamido group), but we also block any substitution *para* to the methyl group. Thus a synthesis of *meta-*

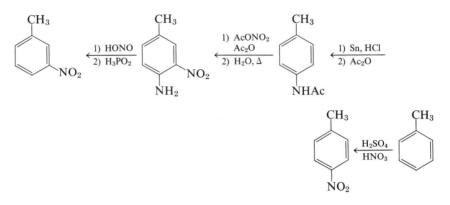

nitrotoluene as outlined here, though long, involves only the manipulation of functionality through a combination of activation and blocking.

The synthesis of *ortho*-nitroaniline represents a similar situation, for here again, the nitration of aniline derivatives occurs predominately in the *para* position. We can solve the functionalization problem by using the

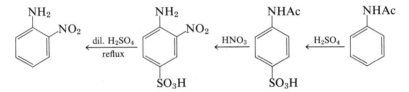

sulfonic acid group both to block the position *para* to the amino group and to deactivate the ring system so that polynitration does not occur. The success of this scheme depends on the reversibility of the sulfonation reaction under aqueous acid conditions.

In both of these cases, we are dealing with the synthesis of disubstituted aromatic systems. The problems become more complex when we consider polycyclic aromatic ring systems or polysubstituted benzene derivatives. Consider, for example, the synthesis of 4,6-dimethoxyphthalaldehydic acid (**85**). Not only is this a tetrasubstituted benzene derivative, but also only

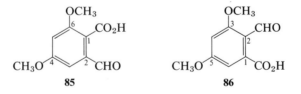

two of the four substituents are alike. In addition to these problems, we are faced with two possible isomers which depend not on substitution but on the oxidation state of the carbon substituents. Thus we must design a synthesis of the acid **85** in contrast to the acid **86**.

The first step in the design phase of this synthesis is to recognize that we can control the problem of the location of the aldehyde function by the reduction of the corresponding phthalic anhydride **87**. Thus, partial

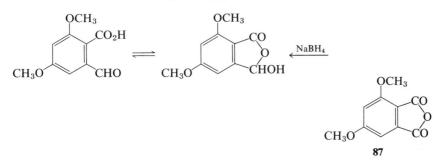

reduction of the anhydride **87** with limited quantities of sodium borohydride will occur at the less-hindered carbonyl group. This means that we have a method for differentiating the two like carbonyl groups of the anhydride **87** by virtue of their differing steric environment, and that the anhydride can serve as a precursor of the desired phthalaldehydic acid **85**. This conclusion also greatly simplifies the synthetic problem associated with the construction of the requisite carbon skeleton, for the ring substituents now consist of two sets of two like groups.

The required carbon skeleton will best be prepared by the addition of the carbon residues to a dimethoxybenzene system. Dimethoxybenzene, as well as many of its derivatives, is commercially available and therefore could serve as starting material. Since there are no reactions available for the simultaneous addition of two adjacent carboxyl groups to a benzene ring, we will have to introduce each carbon residue separately. The two choices for substrates to which the last carboxyl group could be added are

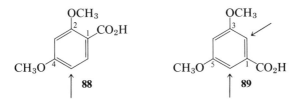

2,4-dimethoxybenzoic acid (**88**) and 3,5-dimethoxybenzoic acid (**89**). The choice here is easy. First, symmetrical compounds are easier to make and hence are attractive starting materials. Such substances will not lead to the generation of structural isomers in subsequent reactions. Therefore, 3,5-dimethoxybenzoic acid (**89**) is the compound of choice on symmetry grounds. Second, an entering substituent in a dimethoxybenzene system will preferentially attack the position *para* to one methoxyl group and *ortho* to the other. In the case of the symmetrical 3,5-dimethoxybenzoic acid (**89**), either of these two available positions will lead to a desired carbon skeleton, whereas 2,4-dimethoxybenzoic acid (**88**) will generate an undesired skeleton in which the carbon residue will enter *meta* to the carboxyl group. Therefore, on both symmetry and electrostatic grounds, our choice of intermediate is 3,5-dimethoxybenzoic acid (**89**).

At this point, the highlights of the synthesis are described. Ethyl 3,5-dimethoxybenzoate (**90**) is commercially available, and we need only to introduce the last carbon residue. Brockmann, Kluge, and Muxfeldt[12a] accomplished this by modification of a very early procedure by Fritsch (1897),[12b] where chloral hydrate is the source of the last carbon atom. Thus, though we introduce two carbon atoms and retain only one, the efficiency of the method is found in the excellent overall yield.

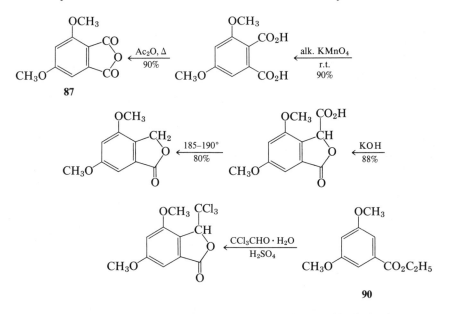

Even though this route results in the contruction of the desired system in good yield, we might consider yet another, shorter approach. 3,5-Dihydroxybenzoic acid (**91**) is also commercially available and undergoes a Gattermann aldehyde synthesis[12c] very efficiently. Use of the Gattermann reaction provides for the introduction of only the one necessary carbon residue, and we no longer need the various stages described before to modify the two-carbon chain. Inasmuch as the yield in the Gattermann reaction is higher when the free phenol is used, we have to add the methylation stage to attain our goal. It is also interesting to note that this

[12] (a) H. Brockmann, F. Kluge, and H. Muxfeldt, *Ber.*, **90**, 2302 (1957). (b) P. Fritsch, *Ann.*, **296**, 344 (1897); (c) W. E. Truce, "The Gattermann Synthesis of Aldehydes," *Organic Reactions*, **9**, 37 (1957).

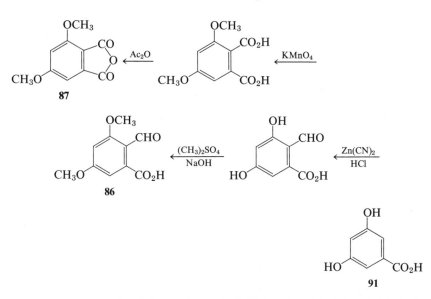

sequence provides both isomeric phthalaldehydic acids **85** and **86,** since the latter is an intermediate in the synthesis of the former.

Here again we see that there is no absolute answer to the synthesis of a specific compound. At least two viable routes are available, and either would prove useful. We should note in this example the effect of steric hindrance in the control of the reduction of the anhydride **87** and the utility of symmetrical synthetic intermediates, for both synthetic plans begin with derivatives of the same symmetrical benzoic acid **89.**

6-METHOXYTRYPTAMINE (92)

Here we consider the synthesis of a polycyclic heteroaromatic material. The most efficient method for the construction of such aromatic systems bearing a carbon side-chain substituent is to make the aromatic nucleus first and then add the side chain. Before this basic scheme may be adopted, however, we must assure ourselves that the side chain may be appended in the desired location on the aromatic system. When considering a new compound, we rely on analogy and organic mechanistic theory. We find that indole is readily acetylated[13] in the 3-position due to the interaction

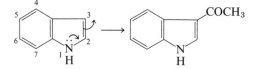

[13] P. L. Julian, E. W. Meyer, and H. C. Printy, "The Chemistry of Indoles," in *Heterocyclic Compounds,* Vol. 3, R. C. Elderfield, Ed. (New York, N.Y.: J. Wiley & Sons, Inc., 1952).

of the nitrogen unshared electron pair with the pyrrole double bond. Thus electrophilic substitution of the indole nucleus will take place in the desired 3-position. A similar effect should predominate in 6-methoxyindole (**93**) even though the methoxy group tends to activate the 2-position toward electrophilic substitution. The 6-methoxy group's effect is diluted by a greater distance.

This analogy settles any question of the proper overall synthetic scheme. The approach should be:

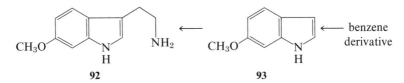

After making this decision by consideration of the desired synthetic objective, we turn our attention to the exact synthetic plan and again find a solution by working backwards. The first stage is the addition of the β-aminoethyl side chain to the 6-methoxyindole nucleus. If we follow the analogy cited before, we will use an acylating agent in the Friedel–Crafts-type reaction as the source of this two-carbon side chain. We must also provide for the incorporation of the primary amino group in this acylating agent; thus both carbons of the side chain precursor must be functionalized. We find such a reagent in oxalyl chloride, in which both carbon atoms are

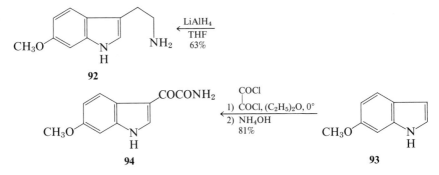

identically functionalized. This is an ideal reagent, for it may serve to acylate both 6-methoxyindole (**93**) and ammonia without concern for selectivity because it is symmetrical. The resulting oxamide **94** is an amide and a 3-indolyl ketone, both of which are readily reduced to saturated carbons by lithium aluminum hydride. Therefore, oxalyl chloride is the reagent of choice for this transformation. In line with the philosophy of preparing the carbon skeleton first and then manipulating the functional groups, we would plan to acylate 6-methoxyindole (**93**) with oxalyl chloride and then treat the resulting acid chloride with ammonia to generate the oxamide (**94**). Finally, reduction with lithium aluminum hydride will accomplish the desired conversion.

The preparation of 6-methoxyindole (**93**) is not as standard, for there are many indole syntheses[13] from which to choose. The approach used in all is, in general, the same. The more stable benzene ring serves as a foundation on which the pyrrole ring is built. For the case at hand, R. B. Woodward and associates[14] chose a rather novel approach in which the pyrrole ring was formed during reduction of the dinitro aromatic system **95**. The key intermediate is 4-methoxy-2-nitrobenzaldehyde (**96**),

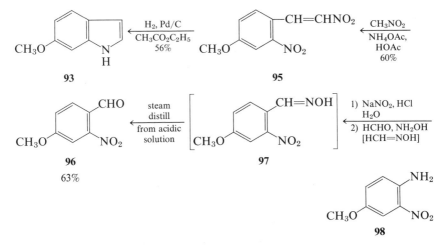

which was prepared through the corresponding aldoxime **97**. Introduction of the formyl derivative into the benzene ring was accomplished by the powerful and generally useful process of displacement of a diazonium salt obtained on diazotization of 2-nitroanisidine (**98**). The amine **98** serves as starting material. This synthesis represents rather specialized chemistry, and its important contribution to the present discussion is as another example of molecular architecture. As unique as the approach may be, it still demonstrates the utility of planning a synthesis from the complex molecule to the simple starting materials. Also apparent is the central role the benzenoid derivatives play as starting materials for synthesis. Here the whole planning process is designed to reduce the 6-methoxyindole (**93**) to a readily available benzenoid component.

2-(3-BUTENYL)-3-METHYLCYCLOHEXENONE (99) AND
4-(3-BUTENYL)-3-METHYLCYCLOHEXENONE (100)

These two compounds represent an interesting example of the variation in synthetic approach necessary to prepare isomeric substances. The two ketones **99** and **100** differ only in the placement of the butenyl side chain. Yet we shall see that each requires a significantly different synthetic

[14] R. B. Woodward, F. E. Bader, H. Bickel, A. J. Frey, and R. W. Kierstead, *Tetrahedron*, **2**, 1 (1958).

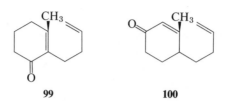

99   **100**

approach. Inasmuch as both cases represent substituted six-membered ring compounds, a logical general approach might be to plan to add the butenyl side chain to a 3-methylcyclohexenone precursor. The position of the side chain attachment dictates the proper precursor.

Consider first the C2-substituted ketone **99.** The position of the butenyl side chain adjacent to the carbonyl group and on the double bond suggests that direct alkylation of 3-methylcyclohexenone should suffice to prepare the compound. In particular, past experience has shown that at

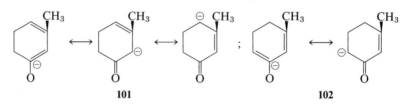

101   **102**

equilibrium the conjugated enolate **101** is more stable than the unconjugated enolate **102;** thus the alkyl residue will be introduced on the desired side of the ketone. We have mentioned such an alkylation before in connection with the conversion of the unsaturated diketone **40** to the keto-

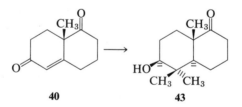

40   **43**

alcohol **43.** In this case, two methyl groups were attached. But when we use the bulkier butenyl bromide necessary here, we can expect the second alkylation to be more difficult. Therefore, monoalkylation of 3-methylcyclohexenone **(104)** should be possible. Under the influence of the reaction conditions, the initially formed β,γ-unsaturated ketone **103** should

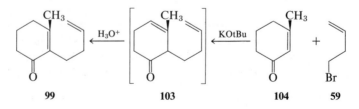

99   **103**   **104**   **59**

also be isomerized to the more stable α,β-unsaturated ketone **99** in which the double bond is now tetrasubstituted as well.

The next step is to consider how the two reactants above might be obtained. We have already described the synthesis of 4-bromo-1-butene (**59**) in connection with the preparation of 5-hexenoic acid (**58**). 3-Methylcyclo-hexenone (**104**) is also readily available by hydrolysis and decarboxylation of 4-carbethoxy-3-methylcyclohexenone (**105**). This ketoester **105** is a vinylogous β-ketoester, which explains the ease of decarboxylation; it is prepared by the interesting reaction between formaldehyde and ethyl acetoacetate. The unsaturated ketoester **105**—Hagemann's ester—is so

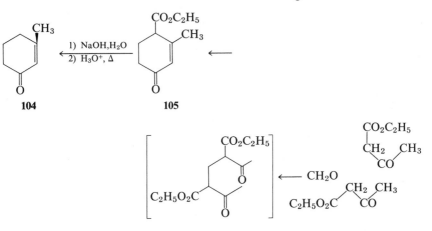

104   105

readily prepared and widely used for the synthesis of 3-methylcyclo-hexenone that it has acquired the name of the investigator who first pre-pared it.[15]

This compound's wide use is not only as a source of 3-methylcyclo-hexenone (**104**) itself, but also because it was found to be easily alkylated adjacent to the ketone group. Hagemann's ester (**105**) is a vinylogous β-ketoester and may be treated in the same fashion as ethyl acetoacetate. Alkylation occurs under the usual basic conditions, and hydrolysis and decarboxylation may be carried out in the standard fashion. Very early it was shown that the alkyl residue was adjacent to the ketone and not to the ester, as might have been the case.

The realization that Hagemann's ester is an intermediate in our synthesis of 3-methylcyclohexenone (**104**) suggests a modification in the synthetic plan. Since the ease of alkylation of the β-ketoester system is usually much greater than that of the direct alkylation of an unactivated ketone, it will behoove us to introduce the butenyl side chain prior to decarboxylation of the ester function. In this manner, W. S. Johnson and

[15] For the best preparative procedure, see L. I. Smith and G. F. Ronault, *J. Am. Chem. Soc.,* **65,** 631 (1943).

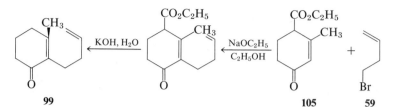

99   **105**   **59**

co-workers[16] were able to realize a 27% overall yield of the desired 2-bute-nylcyclohexenone **99**. The low yield was a result of some difficulty experienced during the hydrolysis and decarboxylation step.

Since we have just shown that Hagemann's ester (**105**) is a precursor of 2-alkyl-3-methylcyclohexenones and *not* of the 4-alkyl isomers, this keto-ester **105** will *not* be useful as an intermediate in the synthesis of the 4-butenyl-3-methylcyclohexenone (**100**). This structural isomer must be prepared by a different synthetic scheme.

The nucleus of the ketone **100** is still the 3-methylcyclohexenone portion. But in view of the preceding synthesis, we must seek another method for its synthesis. One such method is found in the Birch (metal and alcohol in liquid ammonia) reduction[9] of anisole ring systems. The initial reduction product is the dihydroaromatic system which, on hydrolysis with

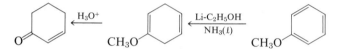

aqueous mineral acid, is converted to the enone. Thus, through such a reduction sequence, an aromatic nucleus can serve as the source of the required cyclohexenone (**100**) if we can devise a synthesis of the appropriate anisole derivative. Again this entails the addition of side chains to an aromatic system, and the scheme is best generated by working backwards.

On the premise that the most efficient synthesis is that in which the carbon skeleton is first constructed and then the desired functional array

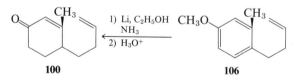

100   **106**

arranged, we would propose the Birch reduction sequence as the last step. Thus, the aromatic system (**106**) is retained until the last possible stage, and it acts as a relatively inert handle while other structural changes occur.

The carbon skeletal construction problem is one of adding the two side chains to the anisole ring. The path of least resistance is the one that

[16] W. S. Johnson, P. J. Neustaedter, and K. K. Schmiegel, *ibid.*, **87**, 5148 (1965).

appends the butenyl side chain *para* to the methoxy group. Our knowledge of aromatic chemistry suggests that substitution *para* to a methoxyl group is a much more efficient process than the introduction of a substituent in the *meta* position. Therefore, the four-carbon chain should be appended to *meta*-cresyl methyl ether (**107**) rather than attempting the methylation of a *para*-butenyl anisole derivative. Not only is this approach attractive from

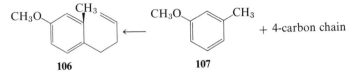

106   107   + 4-carbon chain

a substitution standpoint, but also we recognize that *meta*-cresyl methyl ether (**107**) is an inexpensive commercial substance admirably suited to the role of starting material. Adding the four-carbon unit is the next problem.

This four-carbon side chain is best introduced as a single unit, if possible; as such, it must contain functionality at both ends. One functional group must be suitable for attachment to the aromatic ring, and the other must be either a double bond or a suitable precursor. Such a compound is succinic anhydride (**108**), and W. S. Johnson and associates[16,17] chose this material to accomplish the desired conversion. The succinoylation and reduction sequence is an excellent method for the preparation of γ-aryl-

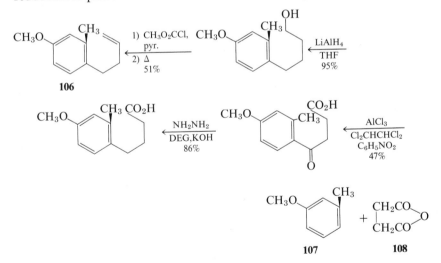

butyric acids, and the yields are generally high. The greater influence of the methoxyl over the methyl group in orienting the site of attachment of

[17] W. S. Johnson, S. Shulman, K. L. Williamson, and R. Pappo, *J. Org. Chem.,* **27,** 2015 (1962).

the succinic acid side chain results in the desired carbon skeleton being the major isomer formed.[17] Transformation of the terminal carboxyl group to the terminal olefin follows standard (but experimentally tedious) procedures.

The foregoing discussion demonstrates how small structural changes can create large variations in synthesis. Analogy to prior art is a very valuable cornerstone of synthetic planning, but unless the structural analogy is almost exact, the conclusion may be misleading. The synthesis of one of the ketones **99** or **100** might seem to be a template for the other, but such is not the case. Nevertheless, following our outlined methods of synthetic planning, one can easily devise adequate routes to both compounds.

4-p-METHOXYPHENYL-4-METHYLCYCLOHEXANONE (109)[18]

Preliminary evaluation of the structure of this ketone **109** reveals two synthetic problems. First, we are faced with a quaternary carbon atom at C-4: we must arrange to construct a cyclohexanone which bears two unlike substituents on one carbon atom. This is difficult to do directly and usually means that, at some stage in the scheme, we must provide functionality adjacent to this carbon so as to effect the substitution desired. Second, we notice that the ketone **109** bears no functionality near the quaternary C-4 position, but only across the ring from it. Some way must therefore be found to incorporate at some stage in the synthetic scheme functionality suitable to facilitate the introduction of the C-4 substituents but also easily removable before the end product is achieved.

Closer examination of this ketone **109** suggests that the major synthetic task will be the construction of the cyclohexanone ring. It is apparent also that the starting material for the synthesis will be a *para*-substituted anisole ring system, since it is unlikely that this aromatic component can either be added at a late synthetic stage or constructed from scratch during the synthesis. If we accept this approach, then the logical place to

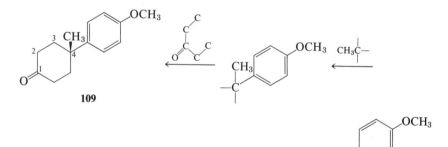

[18] R. J. Ning and R. E. Ireland, unpublished results.

introduce the required methyl substituent is at the stage of the *para*-substitution of the anisole ring system and before construction of the desired cyclohexanone ring. At this stage it will be necessary to have some functional group adjacent to the aromatic ring to arrange for the addition of the rudiments of the new saturated carbon ring. We should plan to make use of this functionality to introduce all the necessary C-4 substituents before it is destroyed during ring construction. By this reasoning, we arrive at the general synthetic scheme outlined earlier, and we are now in a position to devise the actual synthetic details.

As usual, we should approach this planning process by first designing a procedure whereby the final desired ketone may be prepared from a suitable acyclic intermediate. It is again appropriate to consider first a method for the synthesis of the carbon skeleton, and then procedures for manipulation of the functionality present. A particularly powerful method for the construction of six-membered ring systems is the Robinson annelation sequence,[19] in which an aldehyde or ketone is condensed in a two-stage base-catalyzed process with methyl vinyl ketone.

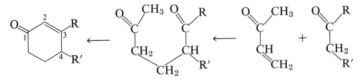

Robinson annelation sequence

There are several modifications of the exact experimental conditions used for this method. But at this stage of the planning process, we should consider only the overall applicability of the sequence to our problem. Notice that a substituent (R′) adjacent to the original ketone or aldehyde component is found on C-4 of the cyclohexenone we generate. Also, there appears to be no reason why two substituents adjacent to this original carbonyl group would alter the course of the reaction sequence, for only one hydrogen α- to the ketone or aldehyde is required. Similarly, if we were to choose an aldehyde for this original carbonyl component, the substituent R would be hydrogen, and the resulting annelated product would simply be a 4-substituted cyclohexenone. To obtain our desired end product, we would merely have to hydrogenate the conjugated double bond. This sequence, therefore, seems admirably suited to our requirements and, as outlined above, requires the propionaldehyde derivative **110** as the starting material. This aldehyde is an excellent substrate for the Robinson annelation process, for the benzylic hydrogen is activated by both the carbonyl group and the aromatic ring. The enhanced acidity of this hydro-

[19] C. D. Gutsche, *op. cit.,* pp. 95–96; H. O. House, *Modern Synthetic Reactions* (New York, N.Y.: W. A. Benjamin, Inc., 1965), pp. 210–212.

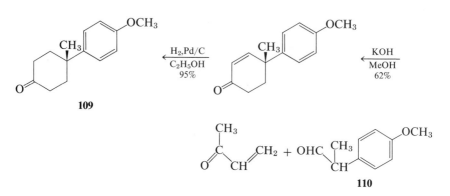

109

110

gen means that mild basic conditions will suffice to cause the initial Michael-type addition to the enone and will thereby obviate the polymeric side products which result when methyl vinyl ketone is treated with strong base.

We turn next to the preparation of the required propionaldehyde derivative **110**. Direct methylation of *p*-methoxyphenylacetaldehyde (**111**) might initially be considered for this stage, but should be rapidly discarded. While monomethylation of such compounds can be accomplished through the use of the pyrrolidine enamine intermediate, the starting aldehyde **111** is a labile substance and is not readily prepared. The addition of the

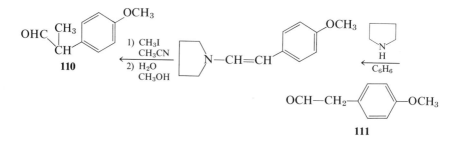

110

111

p-methoxyphenyl group to a propionaldehyde derivative is unacceptable, for there is no satisfactory benzenoid alkylating agent. Even without these restrictions, the most efficient source by far of the desired aldehyde **110** is through introduction of the formyl group in place of a ketone carbonyl of commercially available *p*-methoxyacetophenone (**114**). Two very satisfactory methods are available for this conversion. The ketone **114** may be condensed with methoxymethylenetriphenylphosphorane in a Wittig reaction, and the resulting enol ether **112** hydrolyzed in aqueous mineral acid. The older but equally successful Darzen's reaction sequence may also be employed, wherein the glycidic ester **113** is formed from the ketone **114** and then saponified and decarboxylated. Either route may be used to

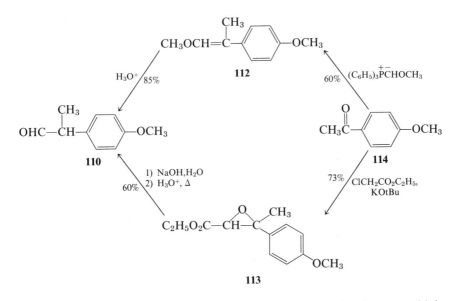

accomplish the desired transformation in approximately the same high yield, and both start from commercially available material.

The foregoing synthesis is an example of the process of planning a synthetic scheme for a structure where the functional array does not readily guide the skeletal planning process. Here, careful analysis of the desired structure and a knowledge of general sequences of several reactions which overcome major structural features combine to make an apparently difficult synthesis routine.

TRANS-9-METHYL-1-DECALONE (115)

The synthesis of this molecule poses both structural and stereochemical problems. The compound in itself is important in the development of synthetic methods in that it is a model for similar necessary transformations in large, more complex systems. For instance, a synthesis[20] of (±)-estrone (26) entails the introduction of a methyl group in the "angular" C-13 position and then contraction of the six-membered D ring to a cyclopentanone. For this conversion, a model compound was sought that would have enough of the features of the steroidal system so that procedures developed on the model would be applicable to the more difficultly prepared tetracyclic system. A very similar synthesis is that of *trans*-9-methyl-1-decalone (115), and this served admirably as a model for the steroid system.

[20] W. S. Johnson, D. K. Banerjee, W. P. Schneider, C. D. Gutsche, W. E. Shelberg, and L. J. Chinn, *J. Am. Chem. Soc.,* **74,** 2832 (1952).

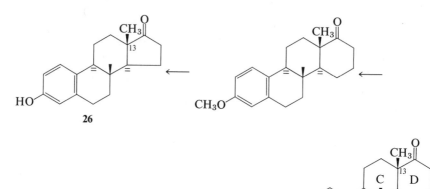

26

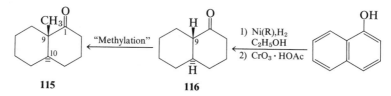

The synthesis of this decalone **115** is influenced by both its structure and the steroid synthesis for which it served as a proving ground. The angular methyl group which makes C-9 a quaternary carbon is adjacent to a ketone and, hence, may be introduced by direct alkylation of the parent 1-decalone (**116**). The latter in turn is readily prepared from commercially available α-naphthol by catalytic hydrogenation over Raney nickel and then

115   **116**

oxidation of the 1-decalol produced with chromic acid-acetic acid. This is an ideal model system, for the compounds are very readily available and are quite similar to the desired steroidal system.

The problems of synthesis come in the "angular methylation" step. Early work[21] showed that the unsymmetrical ketone **116** was exclusively

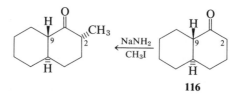

116

methylated on the C-2 carbon rather than the desired C-9 carbon. Thus, to observe the desired angular methylation, it would be necessary to prevent enolization and subsequent methylation at C-2. This situation suggests the use of a "blocking group"—a group that is easily introduced, stable to the proposed reaction conditions, prevents reaction occurring at the site it occupies, and finally, is easily removed or used in further synthetic steps. In the *trans*-9-methyl-1-decalone (**115**) synthesis, the blocking group must prevent enolization of the ketone toward the C-2 carbon and therefore allow enolization toward C-9 so that methylation will then take place in the angular position. Several such groupings have been developed; among them are the N-methylanilinomethylene group,[22] the furfurylidene group,[23] and the *n*-butylthiomethylene group.[24] The last is particularly suitable for the synthesis of *trans*-9-methyl-1-decalone (**115**) because it is readily removed after methylation. The furfurylidene group is useful for the (±)-estrone (**26**) synthesis, for it may be used directly in the D-ring contraction sequence after methylation. The use of the *n*-butylthiomethylene

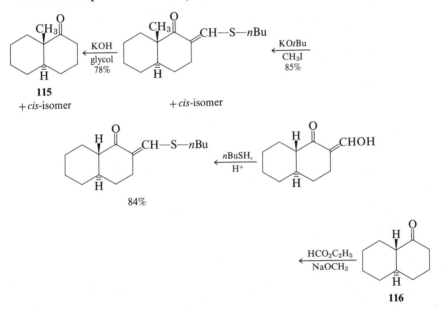

grouping as outlined above is typical of the approach whereby enolization at the 2-position is prevented by replacement of the C-2 hydrogens by a nonenolizable group. Thus the use of a blocking group solves the structural problems of the synthesis; but as indicated, it does not solve the stereochemical problem, for both *cis* and *trans* methylated ketones are formed. Indeed, the stereoisomeric mixture consists of approximately equal portions of each isomer. In practice, the isomers may be separated by careful chromatography; however, no method has yet been devised to

[22] A. J. Birch and R. Robinson, *ibid.*, 501 (1944).

[23] W. S. Johnson, D. S. Allen, Jr., R. R. Hindersinn, G. H. Sausen, and R. Pappo, *J. Am. Chem. Soc.,* **84,** 2181 (1962).

[24] R. E. Ireland and J. A. Marshall, *J. Org. Chem.,* **27,** 1615, 1620 (1962).

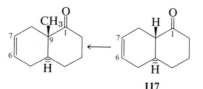

117

accomplish the methylation stereospecifically. W. S. Johnson[22] found that the *trans*-isomer predominates when Δ^6-octalone-1 (**117**) is used in the furfurylidene blocking group sequence; but it is not the sole product. Subtle differences in the character of the transition state for methylation control the stereochemical outcome, and their evaluation is discussed elsewhere.[25]

In this synthesis, we see the utility of adding extra groups to synthetic intermediates in order to accomplish the desired objective. The use of blocking groups is prevalent in the annals of organic synthesis and is not limited to the case at hand. We have already seen amine and carboxylic acid blocking groups used in the Sheehan penicillin (**39**) synthesis, a ketal used to prevent the reaction of a ketone in the synthesis of 6-hydroxy-5,5,9-trimethyl-1-decalone (**43**), and a benzyl ether group used to block the alcohol function in 7-isopropyl-3,7-octadienol-1(**75**). Thus, the blocking group is a particularly useful tool in the synthesis of complex organic molecules and may be used to alter the site of reaction, as above, or to prevent interference by a specific functional group during a synthetic sequence, as in the cases cited earlier.

The efficacy of functional group blocking is nowhere more dramatically demonstrated than in the synthesis of polypeptides. These biologically very important molecules are specific polymers of numerous different amino acids. The problem presented by their synthesis is one of selectivity in construction of the polypeptide chain rather than one of molecular architecture. To prevent the random polymerization of the amino acid components and to gain the selectivity necessary to synthesize a specific polypeptide, the polyfunctional amino acid units must be reduced to monofunctional systems by the addition of blocking groups. Thus the synthetic planning necessary here is one of defining appropriate blocking groups and methods for peptide bond formation, for the polypeptide structure itself defines the sequence. An example of such a synthetic scheme is presented here for the synthesis of L-asparaginyl-L-cysteine (**118**) which employs the N-carbobenzyloxy (Cbz) protecting group and the **p**-nitrophenyl ester activating group, as well as dicyclohexyl carbodiimide (DCC) as a condensing agent. This is only a simple example of the synthesis of a dipeptide; the approach and the auxiliary groups have been used to make much larger molecules. It is apparent that, without these methods, the specific synthesis of even this dipeptide **118,** to say nothing of a decapeptide, would be impossible.

[25] H. O. House, *op. cit.*, pp. 201–204.

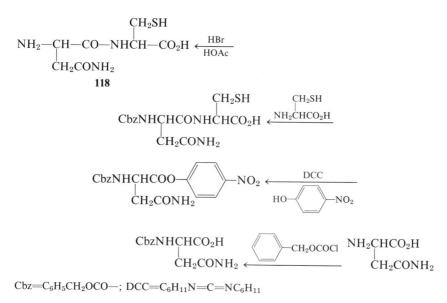

$$CH_2SH$$
$$NH_2\!-\!CH\!-\!CO\!-\!NHCH\!-\!CO_2H \xleftarrow[\text{HOAc}]{\text{HBr}}$$
$$CH_2CONH_2$$
118

$$CH_2SH \qquad CH_2SH$$
$$CbzNHCHCONHCHCO_2H \xleftarrow{NH_2CHCO_2H}$$
$$CH_2CONH_2$$

$$CbzNHCHCOO\!-\!\bigcirc\!-\!NO_2 \xleftarrow{\;DCC\;} HO\!-\!\bigcirc\!-\!NO_2$$
$$CH_2CONH_2$$

$$CbzNHCHCO_2H \quad \bigcirc\!-\!CH_2OCOCl \quad NH_2CHCO_2H$$
$$CH_2CONH_2 \xleftarrow{\hspace{3cm}} CH_2CONH_2$$

$$Cbz = C_6H_5CH_2OCO\!-\!; \quad DCC = C_6H_{11}N\!=\!C\!=\!NC_6H_{11}$$

After this discourse on the utility of blocking groups, let us reconsider the problem of the conversion of 1-decalone (**116**) to *trans*-9-methyl-1-decalone (**115**). This time let's ask the question, "Can we accomplish this conversion without the use of a blocking group?". If so, this would be an ideal solution to the desired structural change, and the answer comes from a close consideration of what is involved in the conversion. The prerequisite for a successful angular methylation of 1-decalone (**116**) is the enolate **120;** yet on direct base-catalyzed enolization the enolate **119**

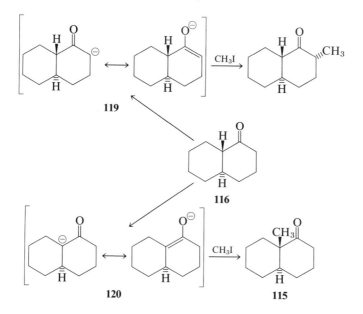

is formed. We solved this problem before by incorporating a blocking group to *prevent* formation of the undesired enolate **119**. We can also solve the problem by choosing suitable derivatives of 1-decalone (**116**) such that *only* enolate **120** is produced. House[26] found that the latter approach is possible if one chooses the correct enol acetate of 1-decalone (**116**) as an intermediate in the following sequence:

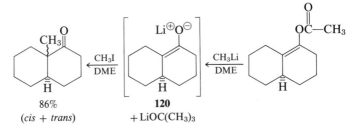

Once formed by the addition of methyl lithium to the ester grouping, the desired enolate **120** will not equilibrate with its isomer and is methylated in the angular position.

Stork[27] also was able to produce this same lithium enolate **120** by the lithium-ammonia reduction of the unsaturated ketone **121**. Again, only

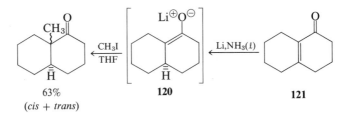

little equilibration with the undesired isomer was observed, and methylation produced the 9-methyl-1-decalones.

These two procedures are very powerful methods for selectively alkylating unsymmetrical ketones. They produce excellent yields of specific products and do not require the additional steps necessary to introduce and remove a blocking group. If for no other reason, the latter consideration makes the procedures attractive because the arithmetic demon is thwarted. The stereochemical outcome in each case, however, is still the same, and a mixture of *cis-* and *trans*-9-methyl-1-decalones is obtained.

Historically, the blocking group was first employed to solve this problem, and in some instances, it is still the best approach. However, the efficiency attendant with the avoidance of a blocking group prompted further consideration of the problem. The rewards of this further work

[26] H. O. House and B. M. Trost, *J. Org. Chem.,* **30,** 2502 (1965).

[27] G. Stork, P. Rosen, N. Goldman, R. V. Coombs, and J. Tsuji, *J. Am. Chem. Soc.,* **87,** 275 (1965).

have been great in that a greater insight into the reaction intermediates resulted, and two new and very valuable procedures were developed. This demonstrates the value of continued curiosity.

6,6-ETHYLENEDIOXY-1(9)-OCTALIN-2-ONE (122)

At the outset, the structure of this ketone **122** suggests the mode of its synthesis. The unsaturated ketone ring fused to another six-membered ring is a classic example of the application of the Robinson annelation[19] sequence. Indeed, this is just the type of compound for which the process

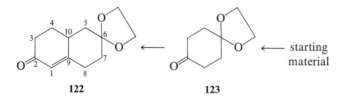

122   **123**

was developed. The general synthetic plan is therefore not hard to devise, and it entails the addition of methyl vinyl ketone or its equivalent to the cyclohexanone derivative **123,** which, in turn, may be prepared from readily available materials. As easy as it is to outline the gross features of this synthesis, the exact details are not so obvious, and they serve to underscore some important synthetic concepts. As usual, we must first consider the last stage of the scheme.

The first step of the annelation sequence is the base-catalyzed Michael-type addition to an α,β-unsaturated ketone. This process requires the preferential conversion of the saturated ketone component to its enolate without the enolization of the α,β-unsaturated ketone addend. To the extent that the latter is enolized, polymerization of this reactant may take place and significantly lower the overall yield of the desired annelation product. Another requirement of the Michael-type addition reaction is that

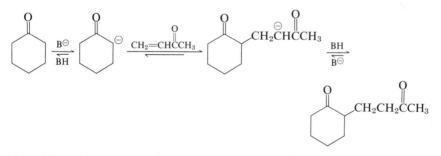

the adduct be protonated, preferably more efficiently than the starting saturated ketone. If the reaction conditions and reactants can be arranged so that these two requirements are met, the otherwise reversible Michael-type addition will be practically irreversible, and high yields of adduct can be expected. One method to accomplish this is to employ a precursor of the α,β-unsaturated ketone portion that will only slowly generate that reactive component under the basic reaction conditions. In this manner, only low concentrations of enone are present at any given time, and polymerization is thus suppressed. Such precursors that have been found effective are the methiodide **124,** formed from the Mannich base reaction product of acetone, and the chloroketone **125,** formed by the Darzen's reaction between acetyl chloride and ethylene. Both β-substituted ketones generate the required methyl vinyl ketone under the basic reaction conditions of the annelation experiment. Both compounds, as well as various homologs

$$(CH_3)_3\overset{\oplus}{N}CH_2CH_2COCH_3 \overset{I^{\ominus}}{} \xleftarrow{CH_3I} (CH_3)_2NCH_2CH_2COCH_3$$

124

$$\xleftarrow[\text{2) } OH^{\ominus}]{\substack{\text{1) } (CH_3)_2NH_2^{\oplus}Cl^{\ominus} \\ \text{HCHO}}} CH_3COCH_3$$

$$ClCH_2CH_2COCH_3 \xleftarrow{SnCl_4} CH_2{=}CH_2 + CH_3COCl$$
125

which retain the β-substituted ketone structural portion, have been successfully used in the annelation process.

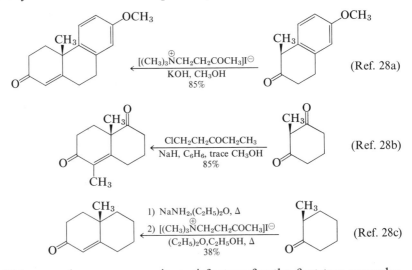

(Ref. 28a)

(Ref. 28b)

(Ref. 28c)

This procedure appears quite satisfactory for the first two examples above; but the yield leaves something to be desired in the last case. Our

[28] (a) F. H. Howell and D. A. H. Taylor, *J. Chem. Soc.,* 1248 (1958); G. Stork, A. Meisels, and J. E. Davies, *J. Am. Chem. Soc.,* **85,** 3419 (1963). (b) R. E. Ireland and R. F. Church, unpublished results. (c) E. C. deFeu, *J. Chem. Soc.* 53 (1937); J. A. Marshall and W. I. Fanta, *J. Org. Chem.,* **29,** 2501 (1964).

present synthetic concern is more closely related to this last example than to the first two; therefore, we might consider this annelation process further in an effort to find a procedure that will result in a higher yield. The difference between the first two cases above and the last one lies in the character of the saturated ketone substrates. In the first two examples, the hydrogen lost during enolization of the saturated ketone is more acidic than that in 2-methylcyclohexanone. In both of these cases, the methyne group is flanked by *two* electron withdrawing groups; hence, enolization toward this position does *not* require such vigorous conditions as does 2-methylcyclohexanone. The milder reaction conditions result in fewer complicating side reactions and, hence, a higher yield of adduct. Therefore, if we are to increase the yield of annelation product, we must increase the acidity of the desired hydrogen adjacent to the saturated ketone, which in turn will facilitate the initial Michael-type addition. Several methods are available to accomplish this.

One elegant method to increase the activity of a carbonyl compound toward Michael-type addition is to employ the pyrrolidine enamine as the reactive intermediate. Though not applicable to the 2-methylcyclohexanone case (the enamine of this ketone substitutes preferentially in the 6-position), this procedure has wide, general use in other cases. For instance, cyclohexanone itself affords a poor yield of annelated material directly, but through the use of the enamine, the yield is increased to 71% of this product.[29] Further examples of the utility of the enamine procedure are plentiful and well described in the research literature.[30]

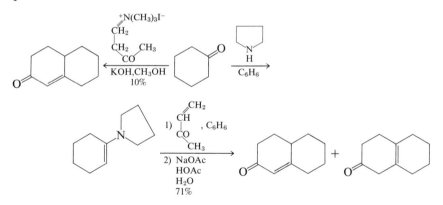

[29] G. Stork, A. Brizzolara, H. Landesman, J. Szmuszkovicz, and R. Terrell, *J. Am. Chem. Soc.,* **85,** 207 (1963); J. Szmuszkovicz in R. A. Raphael, E. C. Taylor, and H. Wynberg (eds.) *Advances in Organic Chemistry: Methods and Results* (New York, N.Y.: Wiley-Interscience, 1963), Vol. 4, pp. 1–113.

[30] H. R. Snyder, L. A. Brooks, and S. H. Shapiro, *Org. Syn. Coll.,* Vol. II, 531 (1943); A. S. Dreiding and A. J. Tomasewski, *J. Am. Chem. Soc.,* **77,** 411 (1955); R. E. Ireland and R. F. Church, unpublished results, after the procedure of A. L. Wilds and R. G. Werth, *J. Org. Chem.,* 1149, 1154 (1952).

Another method by which the yield of an annelation product may be increased is to incorporate an activating group adjacent to the saturated carbonyl. Thus, if we make a β-dicarbonyl compound out of a saturated ketone, the acidity of the α-hydrogen is greatly increased, and the Michael-type addition will take place more efficiently. This process again involves the conversion of the saturated ketone component of the reaction to a more reactive intermediate. For the cyclohexanone case, the desired octalone may be obtained in 50% overall yield by first converting the ketone to 2-carbethoxycyclohexanone.[29]

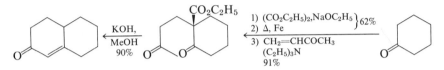

Notice that when the β-ketoester is employed, it is possible to use methyl vinyl ketone itself in the Michael-type addition; triethylamine is a strong enough base to effect the necessary enolization. In this fashion, the Robinson annelation process may be broken down into its two parts and the initial Michael-type adduct isolated. The stronger basic conditions needed to effect the aldol-type condensation are also hydrolytic; therefore, the activating carbethoxy group is removed by hydrolysis and decarboxylation prior to aldol-type condensation. This result means that we can add the carbethoxy group more or less like a catalyst in order to facilitate the overall annelation process.

If desirable, the carbethoxy group may be retained in the cyclized product by employing nonhydrolytic conditions for the aldol-type condensation. Thus, by employing sodium ethoxide as a cyclizing agent, we can effect only the aldol-type condensation and, therefore, realize the synthesis of the ketoester **127**.[31] Conversion of the angular carbethoxy group to a

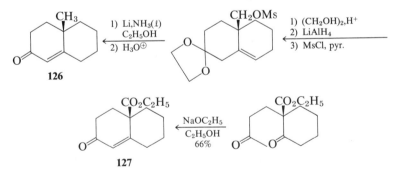

[31] A. S. Dreiding and A. J. Tomasewski, *J. Am. Chem. Soc.*, **77**, 411 (1955); prepared also by W. G. Dauben, R. C. Tweit, and R. L. MacLean, Ref. 31, in 73% yield from 2-carbethoxy-cyclohexanone following the procedure of E. C. Feu, F. J. McQuillin, and R. Robinson, *J. Chem. Soc.*, 53 (1937). See also N. Ferry and F. J. McQuillin, *ibid.*, 103 (1962).

methyl group by the reductive sequence illustrated would then afford another route[32] to the octalone **126,** prepared earlier in poor yield directly from 2-methylcyclohexanone. As we shall see later, this indirect approach to the preparation of similar angularly methylated products has advantages even though it requires more steps.

The foregoing discussion of the Robinson annelation process sets the stage for consideration of the conversion of 4,4-ethylenedioxycyclohexanone **123** to the desired dicyclic ketone ketal **122.** From what was just mentioned, we should not plan to carry out the annelation sequence on the ketone **123** itself; this surely would lead to a low yield. Inasmuch as no angular substituents are present in the desired dicyclic product **122,** either the enamine **129** or the β-ketoester **130** intermediate might at first

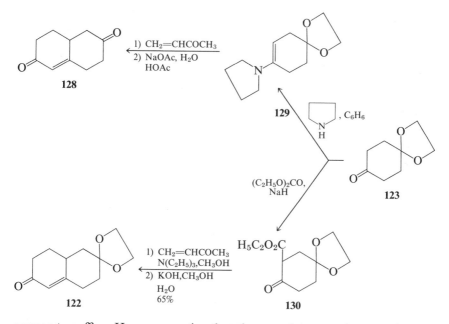

appear to suffice. However, notice that the enamine procedure requires an aqueous, acidic medium for the final stage of the reaction in order to hydrolyze the enamine. Such conditions will certainly hydrolyze the ketal and result in formation of the dione **128.** The β-ketoester approach requires only basic reaction conditions throughout and is therefore more suitable to the synthesis at hand.[28b] It is worthwhile to emphasize that the best way to accomplish the transformation of the ketone ketal **123** to the enone **122** is through the addition of a functional group that does not appear in either the starting material or product. We arrived at this con-

[32] After the procedures of W. G. Dauben, R. C. Tweit, and R. L. MacLean, *J. Am. Chem. Soc.,* **77,** 48 (1955).

clusion by a careful consideration of the prior results of other workers and the mechanistic features of the reaction sequence itself.

We now turn to the construction of the ketone ketal **123**. We are dealing with a saturated six-membered ring, and as we have seen in the past, an efficient source of such systems is aromatic derivatives. We might therefore consider *para*-disubstituted benzene derivatives as suitable starting materials for the synthesis. Catalytic hydrogenation, followed by oxidation, of either benzoquinone or hydroquinone should result in a good

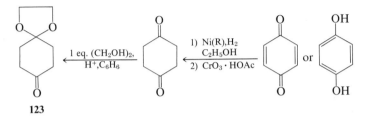

123

yield of cyclohexane-1,4-dione, which in turn may be ketalized with one equivalent of ethylene glycol. The latter transformation is possible only with compounds which contain two identical functional groups and are therefore symmetrical. Even so, this ketalization may be complicated by the formation of a quantity of diketal and by the recovery of the corresponding amount of starting diketone.

Reconsideration of the foregoing general synthetic plan in the light of the potential ketalization problems suggests an alternate approach. If we could form the required ketal prior to the generation of the free ketone, we would obviate this difficulty. Such would be possible if we were to form the cyclohexanone ring by cyclization of a ketalcontaining acyclic precursor. An ideal reaction for this purpose is the Dieckmann cyclization of a seven-carbon diester. If we can obtain the correctly substituted dimethyl

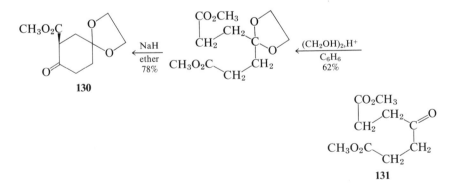

130

131

pimelate, we can not only prepare the desired ketal ketone, but also construct the required β-ketoester directly. For this purpose, the ketal of dimethyl γ-ketopimelate (**131**) is required because, on cyclization[33] with base, the β-ketoester **130** needed for the annelation sequence is produced. The efficiency of this approach is certainly attractive, for it avoids the earlier carbethoxylation step and provides a system which has only one ketone for the ketalization step. In spite of the efficacy of benzenoid derivatives as cyclohexane precursors, this example demonstrates the utility of another approach wherein the functionality present is served more efficiently.

Dimethyl γ-ketopimelate (**131**) is an article of commerce but may also be easily prepared in the laboratory by an efficient and curiously interesting set of reactions. If furfurylacrylic acid (**132**), obtainable in quantity through the Doebner reaction between furfural and malonic acid in pyridine, is treated with hydrogen chloride in methanol, the acid labile

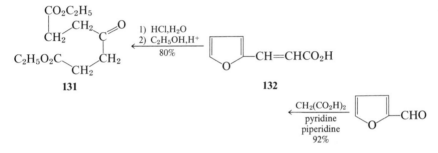

furan ring is cleaved, and the direct result is the formation of dimethyl γ-ketopimelate (**131**) in good yield. This cleavage reaction, first studied by Marckwald[34] in 1887, is an example of the value of the furan nucleus as a source of acyclic γ-oxobutyric acid derivatives. A similar furan cleavage was employed by Robinson[35] for the synthesis of the ketoacid **134** from the furfurylidene ketone **135** in his synthesis of (±)isoequilenin (**133**). Despite the ease and generality of the reaction sequence, it has seen little further use other than that described in these two examples.

We have succeeded in devising an excellent scheme for the synthesis of the dicyclic ketone ketal **122**. The important features are the modification of the Robinson annelation sequence and the use of activating groups to facilitate the reactions. We have seen the utility of activating groups before, particularly in the synthesis of polypeptides, where the way for peptide bond formation was made easier through the use of *p*-nitrophenyl esters and dicyclohexylcarbodiimide. Like blocking groups, activating groups

- 33 R. M. Lukes, G. I. Poos, and L. H. Sarett, *ibid.,* **74,** 1401 (1952); S. Rajagopalan and P. V. A. Raman, *Org. Syn.,* Vol. III, 425 (1955).

34 W. Marckwald, *Ber.,* **20,** 2813 (1887).

35 R. Robinson, *J. Chem. Soc.,* 1390 (1938); A. J. Birch, R. Jaeger, and R. Robinson, *ibid.,* 582 (1945).

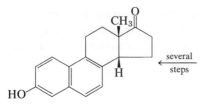

(±)-Isoequilenin (**133**)

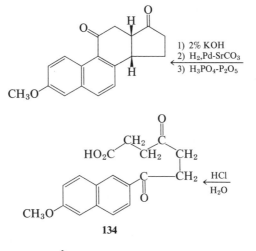

134

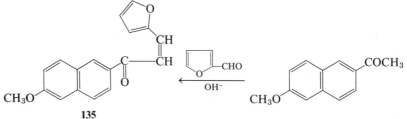

135

facilitate the selective construction of specific structures and have a valuable place in synthesis.

(±)-ZEARALENONE (136)

This material is a naturally occurring macrocyclic lactone. A member of the class of compounds known as the macrolides, which includes such antibiotics as erythromycin, zearalenone (**136**) is one of the structurally less complex members of this class. The total synthesis of zearalenone (**136**)[36] was accomplished by a team of Merck chemists and presents some

[36] D. Taub, N. N. Girotra, R. D. Hoffsommer, C. H. Kuo, H. L. Slates, S. Weber, and N. L. Wendler, *Tetrahedron*, **24**, 2443 (1968).

interesting features. The compound is not complicated by stereochemical problems, for it has only one asymmetric center and a transoid double bond. Structurally, the major feature of the molecule is the large ring lactone, which *a priori* might present some synthetic difficulty. Be that as

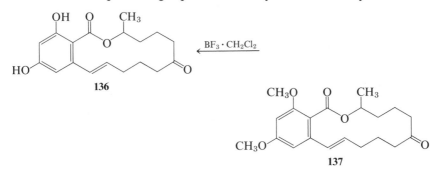

it may, the most obvious and direct route for the synthesis of this lactone **136** is through an intramolecular esterification step. With this approach to guide our planning, we can proceed to the actual design stage.

The first step is to reduce the functionality present in the necessary synthetic intermediates; this can be done efficiently if we work with the phenolic methyl ethers throughout. We can plan to prepare the dimethyl ether **137** of zearalenone by synthesis, and then as a last step cleave the ether groups with boron trifluoride in methylene chloride to free the natural product. This conclusion means that, throughout the synthesis, we can deal with the dimethoxyphenyl system and not be concerned with the problems of handling a phenol.

We are now in a position to consider the formation of the macrocyclic lactone ring. It is already part of our plan to form the ring by attempting the lactonization of the corresponding hydroxy acid **138.** As mentioned, we might anticipate some difficulty in forming the lactone at the expense of the polymeric ester in view of the large ring required. However, notice that the acid is a 2,4,6-trisubstituted benzoic acid and, like mesitoic acid, should be very difficult to esterify under normal conditions due to the steric hindrance offered by the 2- and 6-substituents. Thus, in contrast to the case of an unhindered acid, intermolecular polymeric ester formation should be suppressed here in favor of intramolecular lactone formation.

For the preparation of the hydroxy acid **138,** we might choose one of several routes based on the functionality present in the aliphatic portion of the molecule. But if we remember our logistics, we will choose to construct this hydroxy acid **138** from the two largest synthetic intermediates possible. The largest such fragments related to a functional group are 4,6-dimethoxyphthalaldehydic acid (**85**) and the aliphatic portion of the molecule contained in the phosphonium salt **139.** These two components may be united through use of the Wittig reaction, thereby establishing the

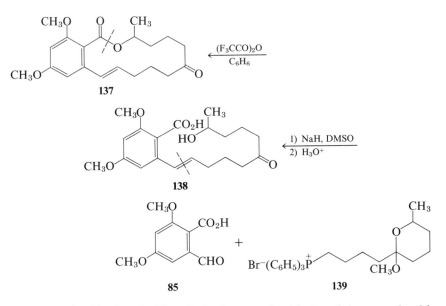

137

138

85   **139**

requisite double bond. The desired *trans* double-bond isomer should predominate from this Wittig reaction but not be the sole product, as some *cis*-isomer can also be expected. Notice that the Wittig reaction may be successfully applied to the aldehydo acid **85** as long as an extra equivalent of base is used to neutralize the carboxylic acid.

Turning to the preparation of the necessary fragments, we notice that we have already discussed the synthesis of 4,6-dimethoxyphthalaldehydic acid (**85**) and therefore have a synthetic plan available. The synthesis of the aliphatic portion of the hydroxyacid **138** represented by the phosphonium salt **139** is another matter. First of all, the choice of this ketal

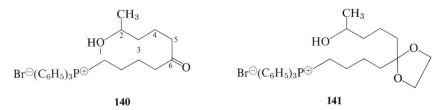

140   **141**

139 is not entirely obvious. The actual aliphatic portion required for the proposed Wittig reaction is the phosphonium salt **140**. However, this material is useless in such a reaction, for the ylide generated would most certainly react intramolecularly with the ketone present. Therefore, to observe the desired transformation, the ketone group must be prevented from interfering. We might accomplish this by masking the ketone in the form of a ketal as in the salt **141**. This blocking group is fine; however, why not make use of the hydroxyl function already present in the molecule as one of the ketal oxygen atoms? This hydroxyl group is situated five carbon atoms from the ketone and can interact to form a six-membered ring. If we provide another oxygen function from a monohydric alcohol such as methanol, we will indeed have the ketone in the form of a stable ketal, and the alcohol function also blocked as an ether. In this manner, we can rationalize the choice of the intramolecular ketal phosphonium salt **139** as the most suitable intermediate for the aliphatic portion of the hydroxy-acid **138**. Notice how the structure of necessary synthetic components and the reactions envisaged serve to define the required intermediates.

We are now ready to devise a scheme for the synthesis of the phosphonium salt **139**. The usual procedure for the preparation of such salts is through the interaction of triphenylphosphine and the corresponding alkyl bromide. The bromide may, in turn, be obtained from the alcohol

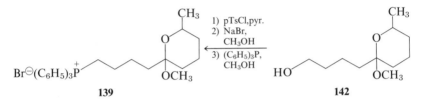

139   **142**

142 by displacement of the derived tosylate with bromide ion. These functional group changes are standard, and while they may be technically difficult, they do not require further comment here.

We can plan functional group transformations for just so long, and then we must again tackle the carbon skeletal construction. We are now at the point where consideration of a suitable precursor for the construction of the alcohol **142** is profitable. Without regard, at first, as to just how the hydroxyl group will arise, but remembering that a functionally substituted chain is necessary, we can approach the carbon skeletal construction of the alcohol **142** by dividing the molecule into the two largest possible fragments. To do this effectively, we can sever the chain at the masked ketone group, whereupon a logical result is the lactone **66** and a four-carbon straight

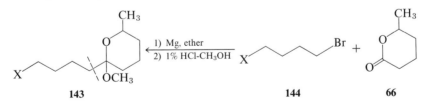

143   **144**   **66**

chain portion. We may now propose that these two fragments be reunited through use of the Grignard reaction between the lactone **66** and the magnesium derivative of an appropriately substituted alkyl bromide **144**. If experimental care is taken during the Grignard addition so as to prevent overreaction and tertiary alcohol formation, one alkyl residue can be

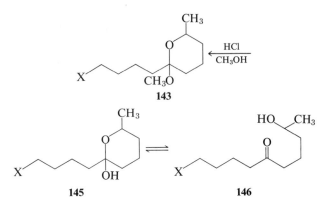

143

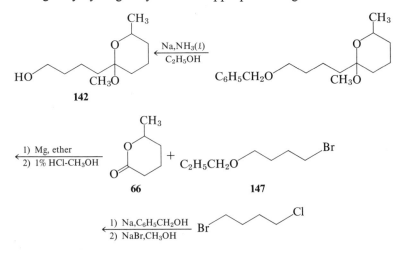

145   **146**

introduced. The resulting hydroxyketone **146** will be in equilibrium with the hemiketal **145** and, on etherification with methanol, will generate the desired ketal **143**.

Two synthetic problems remain: Exactly what substituted alkyl bromide **144** should be chosen, and how may the lactone **66** be prepared? The latter problem was solved earlier in our discussion, and a satisfactory synthetic scheme is therefore available. The choice of the character of the exact alkyl bromide **144** stems from a consideration of the reaction scheme proposed thus far. We obviously *cannot* use 4-bromo-1-butanol, for the active hydrogen on the hydroxyl function is incompatible with the Grignard reagent. We might choose to block this active hydrogen by employing the benzyl ether **147** of 4-bromo-1-butanol and subsequently removing it by hydrogenolysis at the appropriate stage.

This scheme involves a cumbersome preparation of the required benzyl ether **147,** and, in fact, the Merck chemists chose an alternate method for blocking the potential alcohol. A terminal olefin will suffice as a source of the lower homologous oxygenated derivative since, on ozonization, the double bond is cleaved, and an oxygenated derivative results. Thus the double bond is another excellent, stable blocking group for aldehydes, ketones, and, after reduction of the ozonization product, alcohols. The selection of a double bond to mask the terminal alcohol at the intermedi-

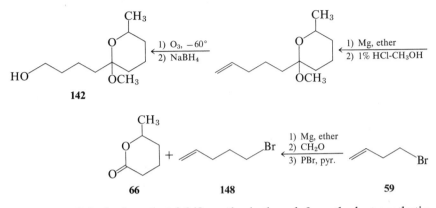

ate stages of the hydroxyketal **142** synthesis then defines the last synthetic intermediate. Thus 5-bromo-1-pentene (**148**) will be required as the alkyl bromide reactant **144** in the Grignard reaction. This pentenyl bromide **148** is readily available from the corresponding butenyl bromide **59** by more or less standard procedures. The latter halide has been used earlier in our discussion of the synthesis of 5-hexenoic acid (**58**).

Thus we have developed a synthetic scheme for the macrolide zearalenone (**136**) which entails twenty-one stages. In spite of the magnitude of the effort and the complexity of the molecule, the synthetic planning is straightforward when taken, step by step, back from the natural product. A great deal of ingenuity and skill on the part of the Merck group was required for this effort, and the success of the synthesis attests to their ability. What we can learn from this discussion is the success of a logical, methodical approach to the planning of such a synthesis.

5

Stereochemistry
Rears Its Ugly Head

It is particularly in the total synthesis of natural products that the problem of stereochemistry becomes significant. The larger, more complex, naturally occurring materials not only possess complicated carbon skeletons and functional arrays, but also are usually one of several possible stereoisomers. Thus an ideal synthetic scheme must furnish the correct stereoisomer as well as the required skeletal and functional group arrangements. To assess the problem of stereoselectivity in synthesis, we will now examine several syntheses wherein the stereochemical aspect is an important added feature.

ISONOOTKATONE

This sesquiterpene occurs together with β-vetivone in the essential oil of vetiver.[1] This oil has a very pleasant odor and is important in the perfume industry. The structures of the two major sesquiterpene components have been the subject of extensive investigation for a number of years, and in 1967, J. A. Marshall and co-workers announced[2] the first total synthesis of the isonootkatone structure 149.

Examination of the structure of isonootkatone (149) reveals the by now familiar α,β-unsaturated ketone system which is available by the Robinson annelation sequence. We are presented with a very similar situation to that encountered in the synthesis of 6,6-ethylenedioxy-1(9)-octalin-2-one (122) discussed earlier.[3] We would therefore predict that the gross structural features of the present molecule may be generated by a similar synthetic scheme. At the outset, then, a suitable general plan would entail the annelation of the β-ketoester 151 with the pentenone 150. The β-ketoester 151 could itself be prepared from the appropriately substituted dimethylpimelate 152 which, in turn, would be synthesized from readily available

---

[1] J. L. Simonsen, *The Terpenes* (London, England: Cambridge University Press, 1949), Vol. 3, p. 224.

[2] J. A. Marshall, H. Faubl, and T. M. Warne, Jr., *Chem. Commun.*, 753 (1967); see also H. C. Odom and A. R. Pinder, *Chem. Commun.*, 26 (1969).

[3] See Chapter 4, pp. 87–94.

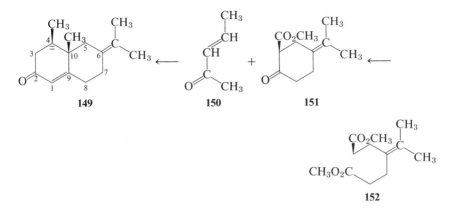

149   150   151

152

starting materials. These gross synthetic features are easy to envisage
after our earlier discussion of the analogous octalone **122** synthesis; but
in contrast to that octalone molecule, the isonootkatone system **149** presents stereochemical problems as well.

The stereochemical problems arise at the annelation stage because,
during the Michael-type addition of the β-ketoester **151** enolate to the
unsaturated ketone **150,** the stereochemical relationship between the
C-4-methyl and the C-10-carbomethoxyl group (the precursor of the
C-10 methyl) is established. This addition could possibly occur so as to
generate one or both of the two stereoisomeric ketoesters **153** and **154.** If

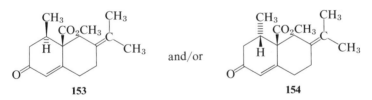

153   and/or   154

only the ketoester **154** were expected, this synthetic scheme would be useless for the isonootkatone **149** synthesis. If a mixture of the two stereoisomers were predicted, the synthesis would be complicated with a tedious
separation problem. We must therefore ascertain the expected stereochemical outcome of this Michael-type addition reaction before we can adopt
this general synthetic approach.

Consideration of the most probable transition state for this addition
reaction led Marshall to predict that the desired stereoisomer **153** would
be formed. The sterically and electronically most favorable arrangement
for the two reactants may be envisaged as the chair-like conformation
depicted below. As is best seen in the Newman projection to the right, this
chair-like arrangement of the transition state places the bulkier vinyl
methyl group outside the cyclohexanone ring system, and only the smaller
vinyl hydrogen lies over this ring.

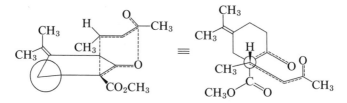

The alternative boat-like transition state conformation leads to the ketoester **154.** This conformation results from rotation of the enone about the carbonyl-olefin single bond, and it requires that the larger methyl group lie over the cyclohexanone ring system. These steric considerations, together with the generally observed[4] preference of such reactions for a chair-like conformation in the transition state, suggest that the major (if not sole) product of this addition will result from this transition state. The stereochemical outcome of this evaluation is that the expected addition product will be the isomer **153** wherein the C-4-methyl and the C-10-carbomethoxyl groups are *cis* to one another. This is also the desired outcome,

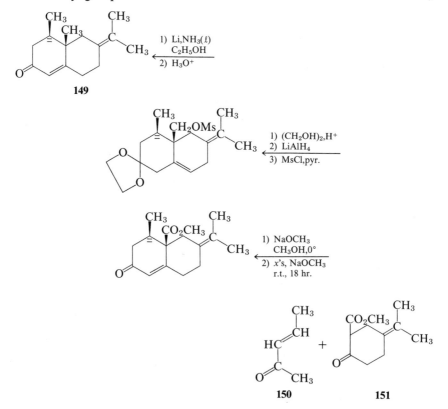

[4] For instance, the Claisen–Cope rearrangement: W. von E. Doering and W. R. Roth, *Tetrahedron,* **18,** 67 (1962).

as the ketoester **153** will indeed lead to the isonootkatone **149** stereochemistry.

Expecting the crucial Robinson annelation sequence to generate the desired stereoisomer, we can consider the actual details of the transformation of the β-ketoester **153** to the natural product **149**. We have previously delineated a sequence for the conversion of the angular carbomethoxyl group to the angular methyl group and would plan to use the same procedure here. Thus the carbomethoxyl group not only facilitates the Michael-type addition to the pentenone **150,** but also serves as the potential angular methyl group. The pentenone **150** chosen for this addition must be the more stable *trans*-isomer in order for our evaluation of the mechanistic details of the annelation process to be valid.

We can now turn to the construction of the β-ketoester **151**, which we already plan to obtain ultimately by Dieckmann-type cyclization of the corresponding pimelate **152**. In contrast to the earlier octalone **122** synthesis, this pimelate **152** is not readily available from furfurylacrylic acid. However, the molecule is symmetrical about the double bond present and, therefore, offers the opportunity to carry out two like reactions to establish the carboxylate-containing side chains. Again approaching this synthesis with the view to combining the largest convenient fragments, we see that the addition of acetic acid residues to a tetramethylethylene derivative will suffice to construct the required seven-carbon diacid skele-

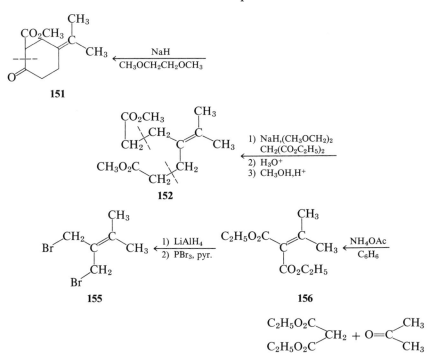

ton. An efficient method for accomplishing this conversion is through alkylation of diethyl malonate and then hydrolysis and decarboxylation. The alkylating agent required for this particular transformation is the dibromide **155.** This dibromide is readily available by standard procedures from the corresponding diester **156.** In turn, Knoevenagel-type condensation of acetone and diethyl malonate produces the required diethyl isopropylidenemalonate (**156**).

This very neat and tidy synthetic plan was realized in practice by the Marshall group, and it led to the synthesis of racemic isonootkatone (**149**). The stereochemical features of the natural product require careful attention to mechanistic detail at the planning stage and careful scrutiny of the experimental results. In the annelation step, these workers found that only the stereoisomer **153** of the crucial bicyclic ketoester was formed. True to prediction, this was the desired isomer, since the ultimate synthetic product was identical to the natural product of known structure. The ability to predict accurately such subtle features of a chemical reaction as the stereochemistry is a powerful result of physical organic chemical investigations; it provides the basis by which we can devise a stereoselective, stereorational synthesis such as this.

α-ONOCERIN

This tetracyclic triterpene would appear to be a logistical challenge as well as a stereochemical problem. The molecule contains thirty carbon atoms and eight asymmetric centers. The latter situation alone means that the natural product is one of 256 enantiomers. Thus, not only will the synthetic effort represent a significant logistical problem, but it also must be stereoselective in order to avoid the generation of numerous unnatural stereoisomers.

As is our practice when approaching any synthesis, we should first consider the gross synthetic outline and then work toward the detailed planning. The striking feature of the α-onocerin (**157**) molecule is the fact that it is a dimer of the dicyclic component **158.** Therefore, if the natural product is cleaved between carbon atoms C-11 and C-12, the resulting halves of the molecule are identical. While this triterpene does not arise

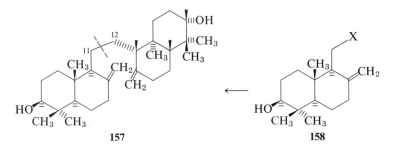

157   158

biogenetically through the coupling of this monomeric derivative, such a scheme is a very attractive approach to the laboratory synthesis, since it also reduces the number of asymmetric centers that must be controlled from 8 to 4. This, then, is an ideal example of synthesis through the union of the two largest possible fragments. However, notice that, due to the asymmetry of the dicyclic component, it is not advisable to dimerize the racemic monomer. Even if we successfully synthesized the correct racemate of the monomer, the dimerization reaction would produce a mixture of two diastereoisomeric onocerin structures—one, the racemic form of the natural product, and the other, a meso compound.

$$d\ell + d\ell \longrightarrow \underbrace{dd + \ell\ell}_{\text{racemic}} \quad + \quad \underbrace{d\ell + \ell d}_{\text{meso}}$$

$$d + d \longrightarrow dd \quad (\alpha\text{-onocerin})$$
$$\ell + \ell \longrightarrow \ell\ell$$

The formation of the unnatural *meso*-compound not only lowers the potential yield of the natural $d\ell$-material, but also necessitates a tedious separation. The problem is obviated if we plan to resolve the monomer and couple the individual enantiomers, for in this fashion, only one enantiomer of the desired natural product is formed in each coupling reaction. Therefore, the gross outline for the synthesis of α-onocerin entails first the stereoselective synthesis of the dicyclic monomer, followed by resolution of this component, and then dimerization of the correct enantiomer to form the natural substance. It is on the basis of this general concept that G. Stork[5] built the first total synthesis of $(+)\alpha$-onocerin (157), and several other workers have successfully followed with alternative approaches to the synthesis of intermediate compounds[6] and different methods for effecting the coupling reaction.[6c]

We can now turn to the exact synthetic planning, and the first problem is to devise a method to effect the coupling of the two dicyclic intermediates. A method is required that will unite two carbon chains without the generation of a new functional group. The bulk of the carbon-to-carbon bond-forming reactions result in the modification of substrate functionality, not the elimination of it entirely. The dimerization of alkyl radicals is an exception. While functionality is required for the selective generation of such radicals, on dimerization, no trace of the initiating groupings remains. Stork[5] chose to form the required radicals by the Kolbe electrolysis of the corresponding carboxylic acid salts, whereby dimerization occurs at the electrode surface and is quite selective. This procedure has been used ex-

---

[5] G. Stork, A. Meisels, and J. E. Davies, *J. Am. Chem. Soc.,* **85,** 3419 (1963).

[6] (a) R. F. Church, R. E. Ireland, and J. A. Marshall, *J. Org. Chem.,* **27,** 1118 (1962). (b) N. Danieli, Y. Mazur, and F. Sondheimer, *Tetrahedron,* **23,** 509 (1967). (c) E. E. van Tamelen, M. A. Schwartz, E. J. Hessler, and A. Storni, *Chem. Commun.,* 409 (1966).

$$CH_3O_2CCH_2CH_2CH_2CO_2H \xrightarrow{\epsilon} [CH_3O_2CCH_2CH_2CH_2]_2$$

159

tensively to prepare long chain diesters as **159**. The high yield of coupled diester precludes any randomization of the radical intermediates. To test the applicability of this electrolysis reaction to the α-onocerin synthesis, the ketoacid **160** was coupled and found to afford a very good yield of

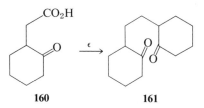

160   **161**

the diketone **161**. Therefore, the scheme for the latter synthetic stages on which Stork decided was the coupling of the ketoacid **165**. It will be noticed that this decision defers until the end of the synthesis the introduction of the exocyclic methylene groups. This timing is reasonable in that the dimeric molecule **164** has one fewer functional group which might complicate the *exo*-methylene formation than does the ketoacid **165**. At first sight, the best method for the introduction of these last two carbon atoms is the Wittig reaction with methylene triphenylphosphorane. Unfortunately, this reagent fails to effect the desired change, probably due to the steric hindrance about the two ketones.

Stork overcame this drawback by taking advantage of the decarboxylation of β,γ-unsaturated acids. Thus, in order to introduce the required exocyclic methylene group, the acetoxyester **162** was formed through the agency of the unhindered ethoxyacetylenic Grignard reagent. On treatment with methanolic sulfuric acid, the initially formed acetylenic alcohol **163** is rearranged to the α,β-unsaturated ester **162**. This alternative to the Reformatsky reaction sequence with zinc and ethyl bromoacetate, and then dehydration, generally occurs in much better overall yield, and the addition step is not nearly as susceptible to steric hindrance.

The decarboxylation step occurs through initial αβ ⟶ βγ-isomerization of the double bond and loss of carbon dioxide through a cyclic mechanism. α-Onocerin (**157**) is known to isomerize under the influence of formic acid to the 8(9)-double bond isomer (β-onocerin); hence, it is reasonable to expect that the corresponding β,γ-unsaturated acid **166** is the reactive intermediate. If this is indeed the case, then the stereochemistry at C-9 of the α,β-unsaturated acid **167** is destroyed by the introduction of the double bond, and it is ultimately the decarboxylation step that establishes the asymmetry at this center. There is reason to believe, however, that the hydrogen at C-9 will be reintroduced in the desired axial (α-) orientation. Decarboxylation of the acid **166** through the cyclic process

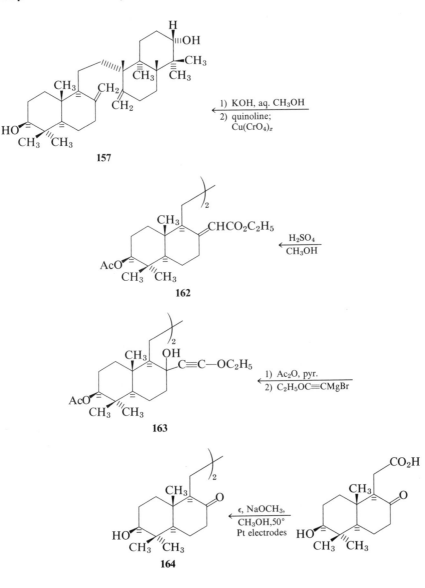

in which the carboxyl hydrogen is introduced axially at C-9 maintains continuous overlap of the developing sp^3-orbital on C-9, with the p-orbitals of the incipient methylene group. This continuous orbital overlap allows greater electronic delocalization in the transition state for decarboxylation; thus, it represents a lower energy pathway than that wherein the hydrogen is introduced at C-9 in an equatorial (β) orientation. The transition state for the latter process does not allow for such electronic delocalization. Thus, in spite of the destruction of the desired orientation of the C-9 sub-

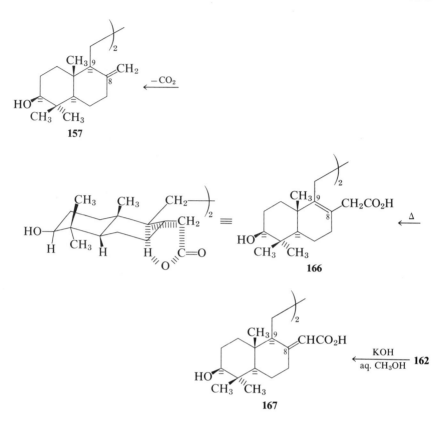

stituents during the isomerization of the α,β-unsaturated acid **167** to its β,γ-isomer **166**, the decarboxylation should still result in the formation of α-onocerin (**157**). In point of fact, Stork found only one isomer formed as a result of this process, and it was the desired natural product. Therefore, if the isomerization does indeed occur in this direction, the decarboxylation must follow the outlined pathway.

We now turn to the synthesis of the ketoacid **165**, the key intermediate in this synthesis. As experimentally taxing as a resolution of optical isomers is, there is little that need be said about this stage of the synthesis. The process was accomplished in the Stork laboratories as follows:

The ($\pm$) ketoacid **165** was first synthesized by Stork[5] and has since been prepared in two other laboratories.[6a,b] The differences in these approaches center around the procedures used to introduce the acetic acid side chain at C-9 and to maintain the ketone at C-8. This variation of synthetic method used to make one key substance attests to the versatility of synthetic organic chemistry, and each approach may profitably be discussed individually.

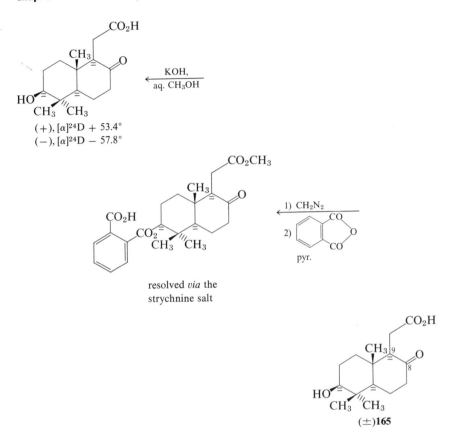

(+), $[\alpha]^{24}D + 53.4°$
(−), $[\alpha]^{24}D − 57.8°$

KOH,
aq. CH_3OH

CO_2CH_3

1) CH_2N_2

2)

pyr.

resolved *via* the
strychnine salt

$(\pm)$**165**

THE STORK APPROACH

We mentioned earlier the utility of the selective cleavage of the cyclo-
hexene ring system as a source of acyclic olefins in connection with the
stereoselective formation of 4-methyl-4-octenedioic acid (**52**). The availability of cyclohexane ring systems from the Diels–Alder reaction as well
as aromatic precursors makes this approach to acyclic molecules particularly attractive. Consideration of the vicinal arrangement of the C-8
ketone and C-9 acetic acid residue in the ketoacid **165** suggests that a similar approach might be suitable here. Since this portion of the ketoacid **165**
requires only four carbon atoms (C-8, 9, 11, and 12), the use of a cyclo-
hexanoid intermediate will necessitate the removal of the two extra carbon
atoms. At least one of these carbons may be eliminated in conjunction
with the introduction of the C-8 ketone. Oxidative cleavage of an
α,β-unsaturated ketone at that position will serve as an excellent method
for the generation of the required carbonyl as well as the removal of one

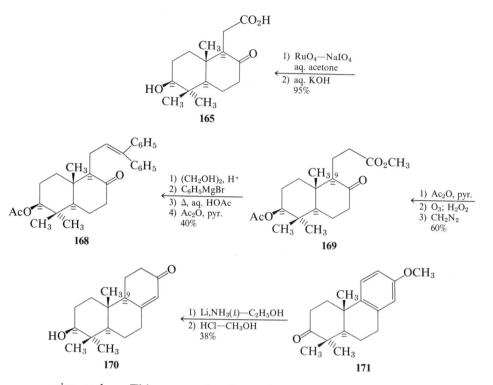

ring carbon. This approach will require that the remaining extra carbon atom be part of the carboxyl containing side chain and, therefore, be eliminated after ring cleavage. No elegant method is available for this transformation; but by the application of modern organic reagents to the classical Barbier–Wieland degradation, lower acid homologs are available in quite satisfactory yields. This approach to the incorporation of the C-8 ketone and C-9 acetic acid residue into a dicyclic ring system allows an aromatic ring to be used as their precursor. Thus, a benzene derivative serves as the source of a cyclohexane intermediate which, on reduction, oxidative cleavage, and finally degradation, will produce the desired δ-ketobutyric acid unit.

The "economy" of the scheme lies in the availability, stability, and latent reactivity of an aromatic ring system. The success of the plan as delineated attests to the power of such an approach, and the concept is well worth emphasis. The only stereochemical point in question in this transformation is that of the C-9 hydrogen. While the final orientation of this hydrogen is not settled until the last step in the α-onocerin (**157**) synthesis, the α-hydrogen configuration is expected in the hydroxyenone **170**, since the enone is formed under conditions which would equilibrate that center to the more stable arrangement. Thus, the *trans-anti* hydrophenanthrene backbone is more stable than the *trans-syn* configuration; hence, the C-9 hydrogen will adopt the α orientation rather than the β orientation.

The next stage in the planning process entails the design of a suitable scheme for the construction of the ketone **171.** Quite adequate procedures for the conversion of the dicyclic enone 9-methyl-5(10)-octalin-1,6-dione **(40)** to the similar 6-hydroxy-5,5,9-trimethyl-1-decalone **(43)** were discussed earlier and may be employed here for the conversion of the tricyclic enone **172** to the necessary saturated ketone **171.** In this case, the stability of the aromatic ring to the various reaction conditions precludes the necessity of adding a blocking group, as in the former example. Again, the stereochemical features of the saturation of the double bond by catalytic hydrogen follow from the earlier arguments.

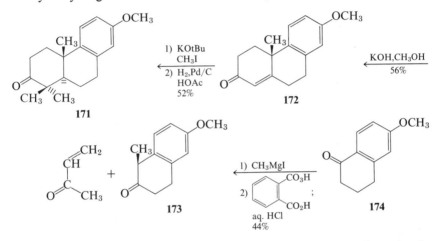

The preparation of the enone **172** has also been discussed previously in connection with the Robinson annelation process. The conditions for the conversion used here by Stork differ from those alluded to earlier only in that methyl vinyl ketone itself was used. The success of the reaction when this labile vinyl ketone is used directly attests to the weakly basic character of the β-tetralone **173** enolate.

Finally, the required 6-methoxy-1-methyl-2-tetralone **(173)** may be prepared from the commercially available 6-methoxy-1-tetralone **(174)** through the Grignard reaction with methyl magnesium iodide and oxidation of the resulting olefin. It is worthwhile to note that the expected tertiary alcohol is not the product realized from this Grignard reaction. Owing to the tertiary and *p*-methoxybenzylic nature of the expected alcohol, we find the ease of dehydration so great that only 6-methoxy-1-methyl-3,4-dihydronaphthalene is formed.

The foregoing reaction scheme was employed by the Stork group for the synthesis of the (±) hydroxyketoacid **165,** and it afforded a 1.1% overall yield. The approach is straightforward, and it accomplishes the desired end result with a minimum of difficulties.

THE SONDHEIMER SYNTHESIS

An alternative approach to the synthesis of the (±)-hydroxyketoacid **165** is to add the C-8 ketone and C-9 acetic acid residue to a preformed dicyclic nucleus. Such a concept entails the positioning of functionality in the dicyclic component at either C-8 or C-9 in order to introduce the side

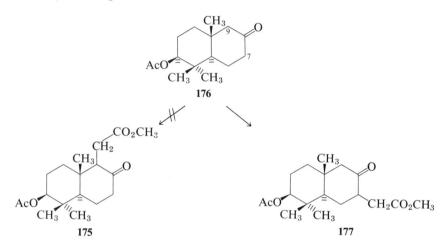

chain. In principle, a ketone at C-8 might serve to activate C-9 toward alkylation with methyl bromoacetate. However, it may be reasoned from steroidal analogies that *trans*-2-decalone derivatives such as the acetoxy-ketone **176** would not enolize toward the C-9 position, but rather toward the C-7 position. Thus, if the acetoxyketone **176** were alkylated without the aid of an activating group at C-9 or a blocking group at C-7, the expected result would be the isomeric acetoxyester **177** and not the desired acetoxyester **175**. A logical method for overcoming this difficulty is to situate a ketone at C-9 so that it may serve the introduction of the acetic acid side chain and then move the oxygen function to the desired C-8 location. Such an approach was planned by F. Sondheimer,[6b] and it depends on an allylic oxidation to accomplish the crucial oxygen movement to C-8. The acetoxyester **179** possesses only one allylic methylene at C-8, and it is just this position in which oxygen is desired. Thus oxidation with selenium dioxide in acetic acid should be specific to this position and should provide a method for the introduction of the required C-8 ketone. In point of fact, the oxidation proceeds well, and the generated alcohol function readily forms the lactone **178** with the unsaturated ester side chain. This is an excellent method for the synthesis of such lactones.

The final transformation to the desired (±)-hydroxyketoacid **165** requires only the internal oxidation-reduction of the alcohol and the double bond. To effect such a conversion, we may take advantage of the $\alpha,\beta \longrightarrow \beta\gamma$ isomerization of the unsaturated ester moiety, for the $\beta\gamma$ isomer is the enol

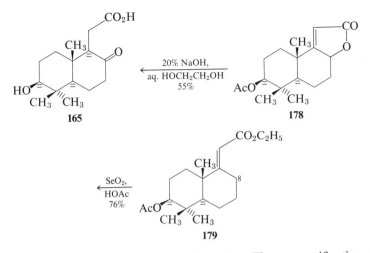

of the C-8 ketone and will rapidly ketonize. Thus saponification with potassium hydroxide serves not only to hydrolyze the C-3 acetoxy group, but also to cleave and rearrange the unsaturated lactone **178**.

Finally, the requisite α,β-unsaturated ester **179** for this sequence is available from the acetoxyketone **180**. Again, we would plan to employ the lithium ethoxyacetylenic reagent in this conversion to ensure a high overall yield and to position the double bond in the α,β-location. The dicyclic acetoxyketone **180** which is the starting material for these reac-

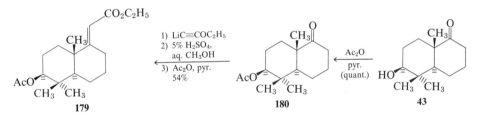

tions is available by acetylation of the hydroxyketone **43**. The synthesis of the latter compound from commercially available substances was discussed earlier.

This scheme represents the more efficient route to the ($\pm$)-hydroxy-ketoacid **165** of the two discussed here. The Sondheimer group was able to realize a 6% overall yield of this key intermediate from 9-methyl-5(10)-octalin-1,6-dione (**40**). The ease of the introduction of the C-8 ketone and the C-9 acetic acid side chain is particularly notable.

SUMMARY

The symmetry of the α-onocerin (**157**) molecule makes the synthetic task somewhat easier, for in one reaction the system may be constructed from its halves. The pedagogic value of the effort has been served by the foregoing discussion; but it is illuminating to observe that these synthetic plans are not the only successful approaches to the onocerin carbon skeleton. As is the case with most complex structures, there are several satisfactory routes to their syntheses. None is more "right" than the others because all will contribute useful information about the character and perfidy of organic reactions. It is through such diverse approaches that the power of modern synthetic organic chemistry is best exemplified. No better case for this tenet is found than in the voluminous synthetic work on the steroid molecule through which many of the major concepts of organic chemistry have either come into focus or been devised.

Another approach to the synthesis of the (±)-hydroxyketoacid **165** was devised by R. E. Ireland.[6a] It is included here in outline form. The key feature where this scheme differs from the foregoing syntheses is in the use of the Claisen rearrangement to introduce the acetic acid side chain. Use of this reaction scheme potentially had a significant advantage

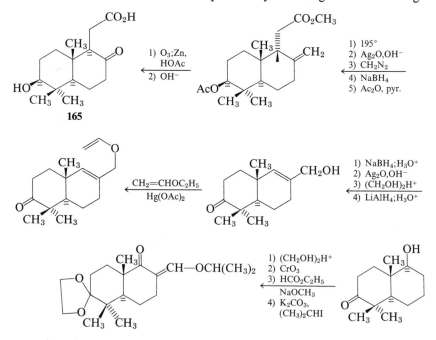

over the other routes because the C-8 methylene group was introduced at the dicyclic stage along with the C-9 acetic acid residue. However, the same reasoning that led Stork[5] to predict the formation of α-onocerin (**157**) on decarboxylation of the diacid **166** applies here as well. The Claisen rearrangement transition state resembles that for the decarboxylation step and results in an α-(axial)-oriented side chain. Epimerization to the desired, more stable β-(equatorial)-acetic acid residue could not be effected without loss of the exocyclic methylene group. Nevertheless, the (±) hydroxy-ketoacid intermediate **165** for the Stork coupling reaction was available in 5% overall yield in sixteen steps.

Finally, in a synthesis patterned after the terpene biogenesis scheme, E. E. van Tamelen[6c] was able to prepare the onocerin skeleton in the form of β-onocerin (**181**). It is particularly pertinent to include this approach to the onocerin synthesis, for it is an example of synthetic planning that follows biogenetic lines. The living systems that prepare the natural products we discuss do so by rational chemical means, albeit through the use of enzymes. Many of these synthetic schemes are amenable

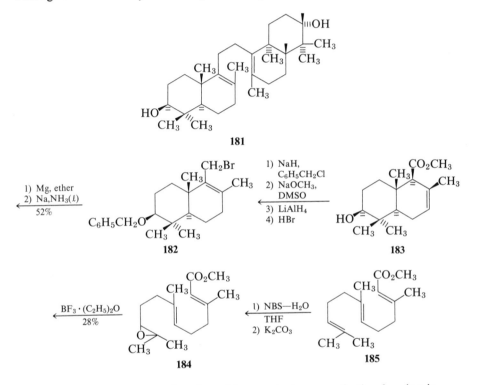

to laboratory realization; therefore, biogenetic-type synthetic planning is another important mode of approaching the construction of naturally occurring substances.

The key biogenetic-type stage in this scheme is the acid-catalyzed cyclization of the polyene oxide **184** to the unsaturated hydroxyester **183**. The latter material was obtained by the selective oxidation of the readily available methyl *trans, trans*-farnesate (**185**) with N-bromosuccinimide in aqueous medium, and then treatment of the resulting bromohydrin with base. It is striking that only one of the three double bonds present in the ester is attacked.

characterization of several of the present structures as potential synthetic intermediates. For one of these structures to serve as a suitable synthetic target, a specific mode for the reunion of the carbons marked with asterisks must be provided. This, in turn, requires the exact designation of functionality at or adjacent to the site of carbon-to-carbon bond formation.

We have already alluded to the possible uses of the Diels–Alder reaction for the conversion of a skeleton such as **192** to the desired carbon framework. This suggests the cyclopentadienyl derivative **193** as an intermediate. The skeleton **191** suggests either the intramolecular alkylation of the bicyclo[2.2.1]heptanone system or the Michael-type addition of the bridged ketone to a side chain enone, as shown in formulation **194**. Another intramolecular Michael-type addition is suggested by the skeleton **187,** the functionalized *cis*-derivative of which is formulated as **195**. While any one of these compounds and others not mentioned specifically warrant consideration as a key intermediate in the longifolene synthesis, Corey[7] chose to investigate the scheme through the enedione **195**. "It is

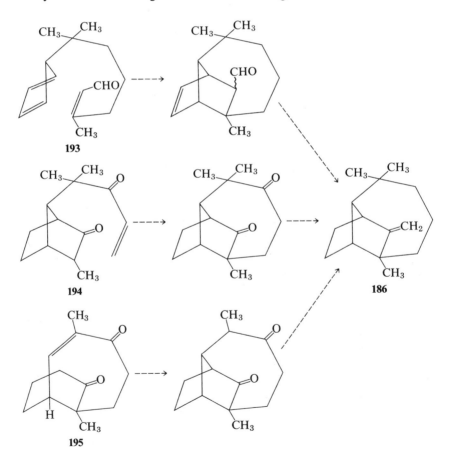

interesting that the choice of a synthetic plan from a broad range of possibilities is very much a function of the methodolgy of synthetic chemistry available at the time, of certain practical considerations such as the availability of the necessary materials and reagents, and of certain subjective judgements relating to the feasibility of key reactions or the existence of alternatives."[7] Thus, a particularly pertinent part of the decision to use this approach was that the enedione **195** represents a simple homodecalin derivative whose synthesis could be expected to be relatively uncomplicated. Equally important was that good precedence for the success of the proposed intramolecular Michael-type addition existed in the conversion of santonin **(196)** to santonic acid **(197)**.[8] Notice that the *cis*-fused isomer

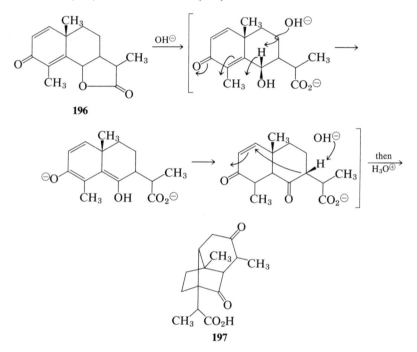

of the homodecalin derivative **195** is required in order to bridge the system in the Michael-type addition. Again, the enedione **195** meets this prerequisite, for the bridgehead hydrogen is labile to base by virtue of its vinylogous activation by the conjugated ketone. As a result, the *cis*- and *trans*-isomers of the enedione **195** will be present in equilibrium concentrations in the base-catalyzed cyclization medium. This situation will then assure the presence of the required *cis*-isomer no matter which isomer is initially introduced.

We turn now to the details of the Corey longifolene **(186)** synthesis and

[8] R. B. Woodward, F. I. Brutschy, and H. Baer, *J. Am. Chem. Soc.*, **70**, 4216 (1948).

see how the schematic general plan discussed above was rendered to practice. As we shall see later, the intermediate enedione (**195**) itself was not isolated *per se;* the β,γ-unsaturated ketal (**199**) was the direct result of synthesis. However, in the light of the known lability of β,γ-unsaturated ketones relative to their α,β-unsaturated isomers under the acidic conditions necessary to hydrolyze the ketal, this modification of the scheme is only of technical significance. The enedione **195** produced in this hydrolytic step was not as readily cyclized as was the santonin (**196**) molecule. However, experimentation led to reaction conditions suitable for the conversion of this material to the bridged dione **198** in 10–20% yield. Even this low yield is remarkable in view of the numerous pre-equilibria necessary to generate the required *cis*-fused enedione, the only moderately activated methylene adjacent to the saturated ketone, and the flexibility of the entire system. In the presence of strong base, the enedione **195** is extensively degraded, and weak base, at 100°, affects little or no change. The success of the intramolecular Michael-type addition, however, is crucial to the synthesis, for the basic carbon framework of longifolene (**186**) is established in this transformation. It was therefore worth the effort to find these rather specific reaction conditions.

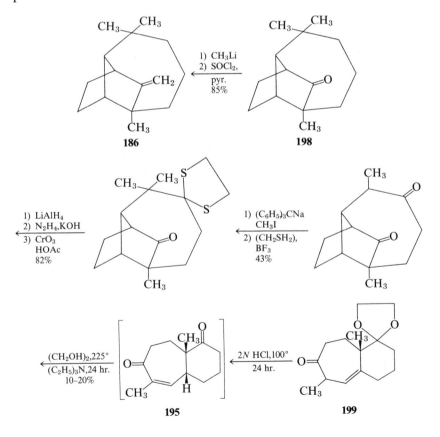

The conversion of the diketone **198** to (±)-longifolene (**186**) is straight-forward, although some experimental difficulties were experienced in the selective removal of the seven-membered ring ketone group. Notice particularly that the methylation stage could be carried out on the tricyclic ketone **198** itself, without fear of methylating the six-membered ring ketone. The only hydrogen adjacent to this carbonyl is at the bridgehead of a bicyclo[2.2.1]heptane system, and enolization toward this position violates Bredt's rule. Therefore, the six-membered ring ketone is inert toward enolization, and consequently, no methylation will take place. While the seven-membered ketone is unsymmetrical and could be methylated on either side of the carbonyl, only the desired *gem*-dimethyl product was observed; therefore, methylene blocking groups were unnecessary.

We now turn to the preparation of the ketone ketal **199** which must serve as the starting material in the crucial intramolecular Michael-type addition. One might consider several methods for the synthesis of such a substance that find analogy in the extensive literature on the synthesis of the lower homologous decalin derivatives. The approach adopted by Corey[7] was to make use of a readily available, suitably functionalized decalin starting material, and to devise a method for the selective expansion of the enone ring. After extensive investigations of various methods for this process, the scheme outlined below was developed wherein the

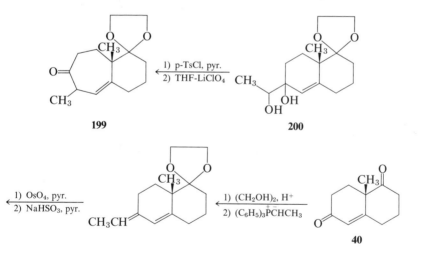

directed pinacol-pinacolone rearrangement serves to enlarge the enone ring system. This is not a trivial conversion, for the success of the entire synthesis depends largely on the synthesis of the ketone ketal **199** as well as on the Michael-type addition. To direct the rearrangement of the diol **200** along the desired pathway, it is necessary to facilitate the ionization of the secondary alcohol; otherwise, the more labile tertiary, allylic alcohol will leave first and result only in an acetyldecalin derivative. For this reason, as well as to prevent ketal hydrolysis, nonacidic reaction conditions are necessary; therefore, the secondary hydroxyl group must be converted to a labile leaving group. This was accomplished by tosylation of the diol **200** which served to generate the secondary tosylate, and then the rearrangement was carried out in an ionizing, nonacidic medium. The migration of the vinyl group in preference to the saturated carbon chain may be anticipated because of the π-electron participation possible in the former grouping. Numerous studies[9] on the migratory aptitudes of various groupings in the pinacol-pinacolone rearrangement substantiate this view.

Finally, generation of the required diol is possible from the readily available octalin-1,6-dione **40** which has also served as a starting material for other syntheses that we have discussed. The selective protection of the saturated carbonyl group is possible by virtue of the deactivation of the unsaturated carbonyl by the adjacent π-electron system.

In practice, this sequence of reactions allowed the formation of the homodecalin ketone ketal **199** in 29% overall yield from the starting enedione **40**. Thus, the only inefficient stage in this ($\pm$)-longifolene (**186**) synthesis is the bridging sequence; but in view of what is structurally accomplished by this Michael-type addition reaction, the low yield is certainly no major drawback. The particular signficance of this synthesis, however, is not so much in the generation of ($\pm$)-longifolene (**186**), but in the important contributions made to the principles and methodology of synthetic organic chemistry. This goal has been admirably achieved.

[9] See E. S. Gould, *Mechanism and Structure in Organic Chemistry*, Holt, Rinehart, and Winston, New York, N.Y. (1959), pp. 607–610.

6

Multistage Synthesis: Logistics and Stereochemistry Combine to Produce Nightmares

As in any field of intellectual endeavor, synthetic organic chemistry has its monumental achievements. In the realm of synthesis, these efforts have classically been the multistage natural product syntheses. More recently, as organic theory has developed, the construction of theoretically challenging molecules has also taken its place among these milestones. However, it is principally in the elaborate syntheses of such molecules as quinine,[1] cortisone,[2] lysergic acid,[3] strychnine,[4] colchicine,[5] chlorophyll,[6] and vitamin B_{12} that the tenets of synthetic design have their foundation. It is therefore appropriate that we deal with the construction of these molecules. To a practitioner of the art of synthesis, every one of the syntheses of the aforementioned molecules is too interesting to neglect, and

[1] R. B. Woodward and W. von E. Doering, *J. Am. Chem. Soc.,* **67,** 860 (1945).

[2] R. B. Woodward, F. Sondheimer, D. Taub, K. Heusler, and W. M. MacLamore, *ibid.,* **74,** 4223 (1952); L. H. Sarett, G. E. Arth, R. M. Lukes, R. E. Beyler, G. I. Poos, W. F. Johns, and J. M. Constantin, *ibid.,* **74,** 4974 (1952).

[3] E. C. Kornfield, E. J. Fornefeld, G. B. Kline, M. J. Mann, D. E. Morrison, R. G. Jones, and R. B. Woodward, *ibid.,* **78,** 3087 (1956).

[4] R. B. Woodward, M. P. Cava, W. D. Ollis, A. Hunger, H. U. Daeniker, and K. Schenker, *ibid.,* **76,** 4749 (1954).

[5] E. E. van Tamelen, T. A. Spencer, Jr., D. S. Allen, Jr., and R. L. Orvis, *Tetrahedron,* **14,** 8 (1961); J. Schreiber, W. Leimgruber, M. Pesaro, P. Schudel, T. Threlfall, and A. Eschenmoser, *Helv. Chim. Acta,* **44,** 540 (1961); R. B. Woodward, *Harvey Lecture Ser.* **59,** 31 (1963–64).

[6] R. B. Woodward, W. A. Ayer, J. M. Beaton, F. Bickelhaupt, R. Bonnett, P. Buchschacher, G. L. Close, H. Dutler, J. Hannah, F. P. Hauck, S. Itô, A. Langemann, E. LeGoff, W. Leimgruber, W. Lwowski, J. Sauer, Z. Valenta, and H. Volz, *J. Am. Chem. Soc.,* **82,** 3800 (1960).

the cruel limitations of time and space work a particular hardship on the author. A decision must nevertheless be made, and rather than do grave injustice to several masterpieces by cursory presentation, it is more in concert with the foregoing discussion to delve into the intricacies of only one. Classical in its timing, yet untarnished by the years in its conception and design, the total synthesis of quinine in 1944 by R. B. Woodward and W. Doering[1] is deserving of our attention.

BACKGROUND

Like most natural products that have been isolated from herbal potions used in early medical history, quinine holds a romantic place in organic chemical endeavors. The alkaloid, together with several congeners, occurs in the bark of the cinchona and remijia species, and preparations from these sources were first introduced into European medical use in 1638 as a cure for malaria. The efficacy of these potions became legendary,[7] but almost two hundred years lapsed before Pelletier and Caventou isolated pure quinine in 1820.

The extensive structural investigations which followed laid the foundation for much of our knowledge of the chemistry of quinoline and pyridine bases. The quinine structure is the result of more than a quarter of a century of degradative work on the part of numerous investigators such as Körner, Baeyer, Koenigs, Skraup, Pasteur, Rabe, and von Miller and Rohde. In 1908, Rabe suggested the structure now known to be correct. The stereochemical conclusions came later, and the full representation of

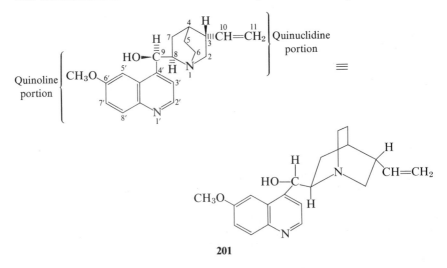

201

[7] For a fascinating account of the background of quinine, see M. B. Kreig, *Green Medicine* (New York, N.Y.: Rand McNally and Co., 1964), pp. 165–206.

the quinine molecule is as shown above.[8] Certain of the reactions observed during the structural work on quinine (201) are particularly pertinent to the synthetic effort.

Quinotoxine (203), the ketone formed from quinine (201) in mildly acidic media, provides an attractive key intermediate—or relay—for the synthetic effort. In his early work, Rabe demonstrated that the alkaloid (201) could be reconstructed from the ketone 203. The cleavage reaction devolves from the presence of the ethanolamine system of the alkaloid in the 4-position of the quinoline nucleus. The acidity of the hydrogens on methyl or methylene groups in the 2- and 4-positions of the pyridine and quinoline nuclei in base-catalyzed condensation reactions is a function of

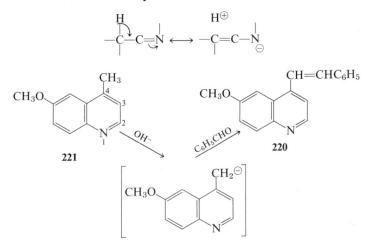

the polarization of the carbon-nitrogen double bond. As a result of this polarization, a hydrogen atom on a carbon adjacent to such a grouping will be relatively easily removed by basic reagents, and the resulting anion may undergo reactions characteristic of an enolate ion. The vinylogous nature of a 4-methyl group in 6-methoxylepidine (221) to the carbon-nitrogen double bond accounts for the acidity of its hydrogens, and, in the presence of base, this compound will readily condense with benzaldehyde to form the styryl derivative 220. The acidity of the hydrogens on such 4-methyl groups, together with the vicinal presence of the quinuclidine nitrogen atom, account for the fragmentation of quinine (201) to quino-toxine (203) under mildly acidic conditions. Both acid and base are required in this transformation, for not only must the hydrogen on the 4-methine be removed by base, but the quinuclidine nitrogen must also be protonated in order to provide an adequate leaving group. Thus the reac-

[8] For a full account of this structural work, see R. B. Turner and R. B. Woodward, *The Chemistry of Cinchona Alkaloids*, in *The Alkaloids*, Vol. III, R. H. F. Manske and H. L. Holmes, editors, (New York, N.Y.: Academic Press, Inc., 1953).

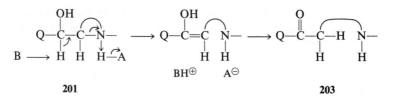

201   203

tion does not take place in either strong acid or strong base, but only in media in which the cooperation of both is possible. The enol that results from the initial fragmentation rapidly ketonizes, and the isolated product is quinotoxine (**203**). This reaction was observed very early in the degradative work, and quinotoxine (**203**) served as an important touchstone in the structure elucidation.

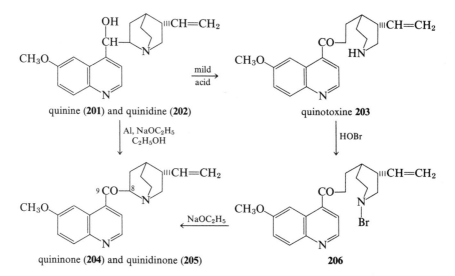

quinine (**201**) and quinidine (**202**)   quinotoxine **203**

quininone (**204**) and quinidinone (**205**)   **206**

Rabe was able to reform the quinuclidine ring system of the alkaloid by the basic dehydrohalogenation of the N-bromo derivative **206**. The latter compound was available from hypobromous acid treatment of quinotoxine (**203**). This ring closure introduces the asymmetric center at C-8; thus, two stereochemical configurations are possible at this position. Inasmuch as the C-8 position is adjacent to the C-9 carbonyl, the spatial orientation about this position may be changed through keto-enol tautomerism. In fact, the C-9 ketone very readily enolizes, and freshly prepared solutions of this ketone slowly mutarotate to an equilibrium mixture of both epimers. One of these two epimers is stereochemically related to quinine (**201**) at C-8 and the other to quinidine (**202**), also a naturally occurring cinchona alkaloid. On oxidation of either alkaloid with benzophenone-potassium t-butoxide (Oppenauer oxidation) or on ring closure of N-bromo-quinotoxine (**206**) with sodium ethoxide, quinidinone (**205**) is isolated; but

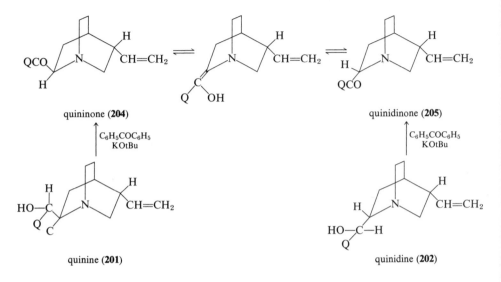

quininone (204)   quinidinone (205)

$C_6H_5COC_6H_5$
KOtBu

$C_6H_5COC_6H_5$
KOtBu

quinine (201)   quinidine (202)

in solution, it slowly changes to an equilibrium mixture of the two ketones
204 and **205**. The fact that only one ketone **205** can be isolated from this
equilibrium mixture is probably a result of its lesser solubility. There is no
compelling thermodynamic reason for the greater stability of quinidinone
(**205**) over quininone (**204**) in the 1-azabicyclo[2.2.2]-octane (quinuclidine)
ring system.

The equilibration of these epimers in solution is fortunate for the synthesis of quinine (**201**) because, on reduction of quinidinone (**205**), both
quinine (**201**) and quinidine (**202**) are formed. Again we have established
a new asymmetric center during this reduction, and each ketone can give
rise to two alcohols. Since both quinidinone (**205**) and quininone (**204**) are
present during the reduction, the reaction product may consist of four
diastereomers. Quinidinone (**205**) may generate quinidine (**202**) and
*epi*quinidine, while quininone (**204**) may produce quinine (**201**) and
*epi*quinine. The *epi*-compounds are indeed present as minor by-products
in the aluminum-ethanol-ethoxide reduction product; but the two naturally
occurring alkaloids **201** and **202** may readily be isolated from this mixture.
Thus this sequence of reactions is a suitable route for the conversion of
quinotoxine (**203**) through quinidinone (**205**) to the desired alkaloid
quinine (**201**). The total synthesis of the alkaloid then requires the synthesis
of quinotoxine (**203**).

QUINOTOXINE (203)

One further set of degradative experiments provides the necessary
components for the construction of quinotoxine (**203**). In 1909, Rabe
showed that quininone (**204**) is cleaved by treatment with amyl nitrite and

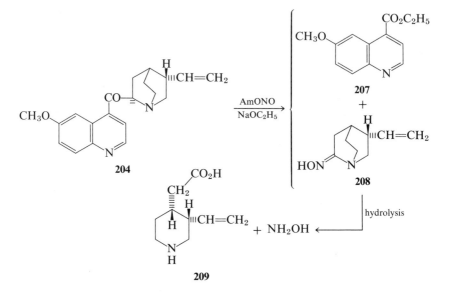

204

207

+

208

hydrolysis

+ NH₂OH ←

209

sodium ethoxide to ethyl quininate (**207**) and an oximino compound **208**, which was readily hydrolyzed to meroquinine (**209**) and hydroxylamine. These reactions are intriguing, and particularly, the amyl nitrite-ethoxide cleavage reaction will become important in the sequel. The formation of the intermediate α-nitrosoketone **210** is analogous to the Claisen–Schmidt condensation between a ketone and an ester; the cleavage reaction is similar to the strong base cleavage of β-diketones and β-keto esters. This nitrite ester condensation is general, and the cleavage reaction is particularly efficient for the fragmentation of ketones which enolize toward a methine position.

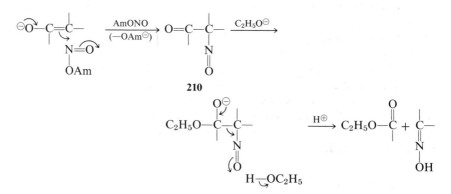

210

Also striking is the ready hydrolysis of the oximino compound **208**. This material and the corresponding lactam (**212**) are significantly more easily hydrolyzed than one would expect of an ordinary amide **211**. While mechanistically the hydrolyses of all three linkages are the same, the explanation for the enhanced reactivity of the former two toward base is found in their bicyclic structure. In ordinary amides **211**, the normal reac-

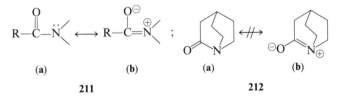

tivity of the carbonyl toward nucleophiles is greatly reduced by delocalization of the carbonyl carbon's partial positive charge over the carbon-nitrogen bond. Thus, an ordinary amide is a resonance hybrid of the canonical forms **211(a)** and **211(b)** in which there is significant double-bond character to the carbon-nitrogen bond. This even manifests itself in restricted rotation about this bond, and the effect is observable in the nuclear magnetic resonance spectra of such common amides as the peptides. The bicyclic character of the quinuclidine nucleus, however, precludes any contribution to such a resonance picture by the carbon-nitrogen double-bond form **212(b)**. This form violates Bredt's rule and would require such high energy as to have no stabilizing effect. The consequence is that the lactam carbonyl of such a bicyclic system retains its additive capacity and behaves more like a ketone than an amide. Such lactams then are readily cleaved by base to their open chain precursors.

Ethyl quinate (**207**) and meroquinine (**209**) not only assisted in the structure work but also were important intermediates in the synthetic effort. The quinotoxine (**203**) molecule has been degraded into the two largest reasonable fragments, and the recombination of these in some fashion follows our previously described tenets of synthetic planning. A very significant contribution to this effort was made by Rabe in 1913 when he showed that aliphatic esters condensed smoothly with ethyl quinoline-4-carboxylate (**215**). The derived β-keto esters **214** were readily hydrolyzed and decarboxylated, and they gave the corresponding quinoline-4-ketones

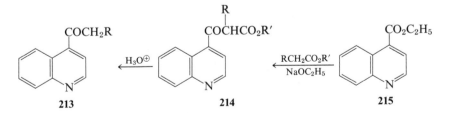

213. Using this sequence, Rabe was able to prepare dihydroquinotoxine (**216**) from ethyl quinate (**207**) and N-benzoylhomocincholoipon ethyl ester (**217**). The latter compound is the ester of the homolog of the dihydro

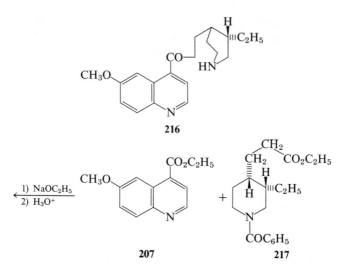

216

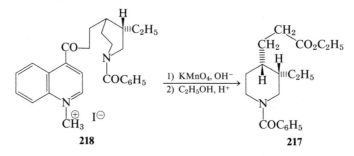

207   **217**

derivative of meroquinine (**209**) (cincholoipon); it is available by alkaline permanganate oxidation of N-1-benzoyldihydrocinchotoxine-N-1'-methi-odide (**218**) and esterification of the resulting acid. This toxine **218** is related to the companion alkaloid cinchonine which differs from quinine

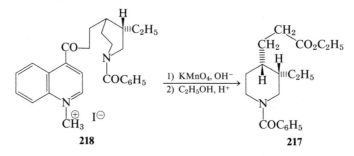

218   **217**

(**201**) only in the absence of the 6'-methoxyl function in the quinoline portion and the stereochemistry at the C-8 position. Subsequently, both ethyl quininate and ($\pm$)-homocholoipon became available through synthesis. In 1931, Rabe completed the total synthesis of dihydroquinine itself by resolution of the ethyl homocholoiponate through the (+)-tartrate and conversion to the optically active dihydro alkaloid by a process similar to that discussed here. It therefore remained for the Rabe ester condensation to be applied to ethyl quininate (**207**) and ethyl N-benzoylhomomero-quininate (**219**) in order to effect the reconstitution of quinotoxine (**203**), and thereby quinine (**201**), from these two fragments. This partial synthesis was carried out by Prelog in 1943, using ethyl N-benzoylhomomero-quininate (**219**) derived from degradation of natural material, and by Woodward and Doering[1] in 1944, employing the racemic ester from total synthesis. The ($\pm$)-quinotoxine (**203**) prepared by total synthesis was

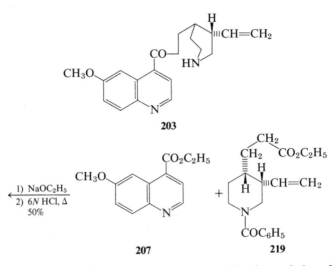

203

207   **219**

resolved through the dibenzoyl-(+)-tartaric acid salts and therefore provides the link between the partial and total syntheses of quinine (**201**). Thus the work of Rabe particularly provided the guidelines for the later stages of the quinine (**201**) synthesis, and it remained to elaborate synthetic schemes for the two major fragments, ethyl quininate (**207**) and ethyl N-benzoylhomomeroquininate (**219**).

ETHYL QUININATE (207)

This aromatic portion of the quinine (**201**) molecule was isolated early in the degradative work. When the time came for its incorporation in a synthetic scheme, several syntheses were already available. Indeed, a great deal of heterocyclic chemistry[9] evolved from this work, and the well-known Skraup quinoline synthesis is only one example. Rather than discuss the first quininic acid synthesis, we will elaborate a more recent method which, consequently, is more efficient.

As mentioned earlier, 4-methylpyridines and 4-methylquinolines are capable of condensation with benzaldehyde because of the methyl activation by the carbon-nitrogen double bond. By analogy to similar benzene derivatives, we would expect the stryryl derivative that results to be susceptible to oxidation and to generate the corresponding carboxylic acid. This analysis suggests that 6-methoxylepidine (**221**) will serve as a suitable intermediate for the ethyl quininate (**207**) synthesis. Indeed, the foregoing reaction sequence provides a very efficient route for this conversion.

Again drawing on our earlier discussions, we would suggest that the synthesis of 6-methoxylepidine (**221**) may best be accomplished through

[9] E. C. Taylor, *Heterocyclic Chemistry* in Foundations of Modern Organic Chemistry Series, Prentice-Hall, Inc., Englewood Cliffs, N.J., 1968.

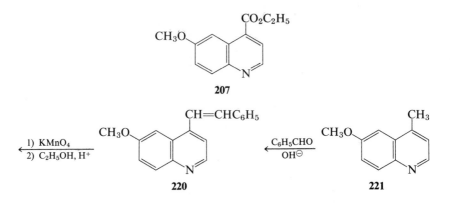

207

220   221

the addition of the nitrogen-containing ring to a benzenoid precursor. A modification of the Skraup synthesis suffices, for the attachment of a sub-stituted four-carbon chain to an aniline derivative, followed by cyclization, should establish the ring system. There are several methods for accomplish-ing this transformation; one of the efficient procedures is shown here. The carbon skeleton is first established by Friedel–Crafts-type cyclodehydra-tion of the amide **223** in 90% sulfuric acid. The resulting quinolone **222** is then converted to the desired 6-methoxylepidine (**221**) by chlorination and

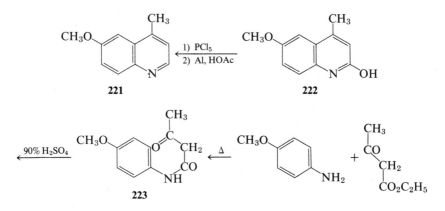

221   222

223

then reduction. Again the sequence is very efficient and provides for a 60% overall yield of 6-methoxylepidine (**221**) from commercially available *p*-anisidine and ethyl acetoacetate.

ETHYL N-BENZOYLHOMOMEROQUININATE (219)

The availability of ethyl quininate (**207**) together with suitable syn-thetic procedures for the formation of quinotoxine (**203**) and its conver-sion to quinine (**201**) leaves only the synthesis of ethyl N-benzoylhomo-meroquininate (**219**) yet to be accomplished. This significant achievement reported by Woodward and Doering[1] in 1944 completed the first total thesis of quinine (**201**).

The structural problems associated with the synthesis of the homo-meroquinine (**224**) molecule relate to the incorporation of the appropriately substituted side chains. The stereochemical challenge pertains to the incorporation of these two side chains in a *cis*-arrangement. The two problems are, of course, related to one another, and a suitable solution to both should be available in one concept.

We can best approach these synthetic problems by first considering the construction of the required carbon skeleton. One can look upon homo-meroquinine (**224**) as a 3,4-disubstituted piperidine, wherein the vinyl and the propionic acid residues are implicitly considered to be appended to the nitrogen-containing ring. Such a viewpoint conjures up a synthetic plan in which these two side chains are added to a preformed piperidine ring system. Such an approach entails a piecemeal method of synthesis and is complicated by the functionality required throughout the scheme. Stereochemical problems can also be anticipated, for there will be little or no control of the steric relationship between the two residues.

A more rewarding yet less obvious view of the homomeroquinine (**224**) molecule is found when it is considered as a 4,5-disubstituted-6-heptenoic acid, wherein the piperidine ring system becomes the appendage on a straight carbon chain. Such a viewpoint emphasizes the presence of the

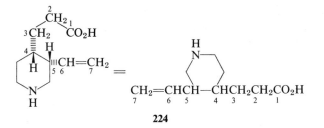

224

seven-carbon straight chain. At first, this may not appear to offer an advantage in the planning of the synthesis, for the addition of a piperidine ring system to such a carbon chain is not an obviously easy task. However, we mentioned earlier the efficiency and economy that attend the synthesis of similar acyclic molecules by the cleavage of cyclic intermediates, and this 6-heptenoic acid derivative is no exception. This straight carbon chain bears functionality at both the 1- and 6-positions, which suggests that it might arise from the cleavage of a cyclohexane ring system. The cyclohexane ring would be preferred over the potentially possible cycloheptane ring (the 7-position of the straight chain is functionalized too) since the former may find its roots in a readily available aromatic ring. This reasoning leads to consideration of the perhydroisoquinoline ring system as a suitable precursor of the desired acid **224**. More specifically, the ketone

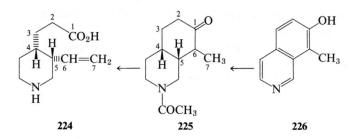

224   225   226

225 would serve particularly well, for oxidative cleavage of the bond between C-1 and C-6 will generate an acid at C-1 and a carbonyl derivative at C-6. The latter can serve as a source of the vinyl group through reduction and elimination. Equally attractive is the prospect that the saturated ketone **225** should be available by reduction of the aromatic isoquinoline system **226**.

These conclusions arise from attacking the synthesis of the homomeroquinine (**224**) molecule as a substituted straight chain carboxylic acid derivative rather than as a piperidine derivative. The ultimate success of the plan points up the importance of considering all aspects of the desired molecular framework. All too often, the most convenient way to draw a molecule on paper belies the most efficient synthetic approach; it is necessary to remember that organic molecules are three-dimensional entities with many faces.

We now turn from the gross conception of the homomeroquinine (**224**) synthesis to the exact details worked out by Woodward and Doering.[1] As incisive as this general plan may be, the selection of the specific reactions that spell success requires an equal amount of ingenuity. Particularly important in this regard is the process whereby the dicyclic ketone **225** is

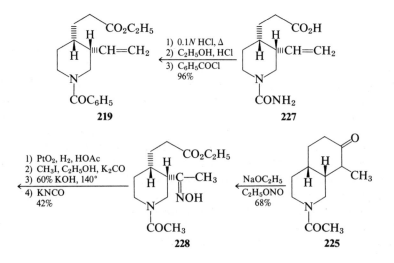

219   227

228   225

cleaved and the desired side chains are formed. In addition to the cleavage of the ketone ring between the carbonyl group and the adjacent methyl-bearing carbon, this key reaction must also provide a product that is suitable for conversion of the two-carbon side chain to a vinyl group. The conversion must also result in no stereochemical equilibration of the side chain configurations. This last condition is particularly demanding, for if the ring cleavage reaction were to generate, for instance, an acetyl group-

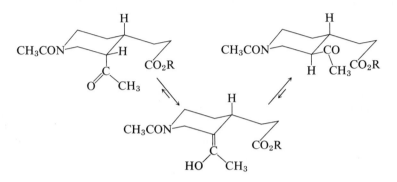

ing, the adjacent ring proton would be labile to acid or base-catalyzed enolization, and the less stable *cis*-arrangement would be converted to the more stable *trans* side-chain configuration. This epimerization would result in the ultimate formation of a quinine-like structure which is epimeric with the natural product at C-3—the carbon bearing the vinyl group. Of course, this cannot be tolerated. So the reaction chosen to cleave the dicyclic ketone **225,** as well as those used to convert the cleavage product to the homomeroquinine (**224**) structure, must provide no opportunity for this type of transformation.

The reaction chosen by Woodward and Doering to accomplish this delicate stereochemical and structural conversion was the nitrite ester cleavage reaction used earlier for the cleavage of quinidinone (**205**) by Rabe. Structurally, this nitrite ester cleavage reaction is ideal, for the ketone carbonyl becomes an ester function, and the adjacent carbon is oxidized to an oxime. The resulting oximino ester **228,** then, is ready for reduction to a primary amine and subsequent base-induced elimination to the desired vinyl group. Such basic (E_2) elimination reactions are known to generate preferentially the less-substituted olefins (Hofmann rule) by removal of a hydrogen from the less-substituted carbon (CH_3 here). Therefore, this process can be expected not only to result in formation of the vinyl group, but also to leave the configuration of this two-carbon side chain undisturbed.

As shown before, this process proved quite satisfactory and resulted in an efficient conversion of the dicyclic ketone **225** through the oximino ester **228** and ($\pm$)-N-uramidohomomero quinine (**227**) to the desired ethyl

(±)-N-benzoylhomomeroquininate (**219**). The latter homomeroquinine derivative is, of course, a necessary component of the condensation reaction that forms (±)-quinotoxine (**203**). The nitrite ester cleavage reaction is also ideally suited to the strict stereochemical requirements of this conversion. Even though an acetyl derivative is generated as the two-carbon side chain, the oximino ester **228** can be expected to be quite resistant to epimerization, even under the basic conditions of the reaction. The most acidic proton in an oxime is that on the oxygen atom. Under the basic

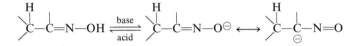

conditions of the cleavage reaction, the product is the oxime salt, and the presence of the resonance stabilized anion in the molecule places a high barrier in the way of proton release from the adjacent asymmetric ring carbon. Therefore, this salt formation provides no suitable pathway for epimerization, and it accounts for the preservation of the desired *cis*-relationship between the two side chains even though the *trans*-configuration is more stable. Thus, on both structural and stereochemical grounds, the choice of this nitrite ester cleavage reaction to effect the conversion of the dicyclic ketone **225** to the oximino ester **228** is central to the ultimate success of the synthesis and demonstrates the keen insight of the investigators.

We must now consider an appropriate scheme for the synthesis of the dicyclic ketone **225**. As mentioned earlier, an attractive feature of this molecule is that it should be available from an aromatic precursor. The

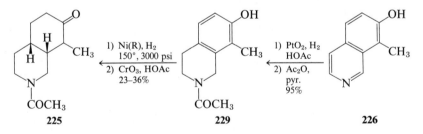

hydroxy*iso*quinoline **226** is the logical choice for this purpose because, on catalytic hydrogenation, and then oxidation, the desired dicyclic ketone **225** should result. In fact, the reduction process had to be carried out in two stages, since the nitrogen-containing ring was reduced first, and the resulting amine poisoned the catalyst. If reduction was forced at this stage, hydrogenolysis of the oxygen function was observed. However, the two-stage approach in which the basic amine generated in the first step was blocked as the acetamide **229,** and the benzenoid ring was then reduced under more vigorous conditions, satisfactorily overcame these problems.

The dicyclic ketone **225** formed by oxidation of the reduction product is a mixture of *cis-* and *trans-*fused isomers. Hydrogenation of the benzene ring apparently does not take place stereoselectively; it is probably the result of sequential absorption of hydrogen from the catalyst, coupled with double-bond isomerism at intermediate, partially reduced states. It was possible, however, to isolate the desired *cis-*isomer from the reduction-oxidation mixture, and thus the synthetic scheme was preserved. It is important to notice that it is the *cis-*isomer of this dicyclic ketone **225** that is stereochemically required for the homomeroquinine (**224**) synthesis; it is at this catalytic reduction stage that this stereochemical feature is established. Thus the carbon skeleton and stereochemistry of homomeroquinine (**224**) are embodied in the dicyclic ketone **225,** and further transformations only reorganize the functionality and bonding.

The synthetic scheme has now been reduced to the construction of an aromatic species—a synthesis that is usually more readily accomplished than that of similar aliphatic species. The approach to this hydroxy*iso*quinoline derivative **226** is similar to that used for previously discussed benzoheterocycles in that the nitrogen-containing ring is added to a benzene derivative. Rather than belabor the actual construction of an *iso*quinoline nucleus for which there are several standard procedures, let us concentrate on the conception of the general synthetic approach. The hydroxy*iso*quinoline **226** represents a benzene ring to which a pyridine

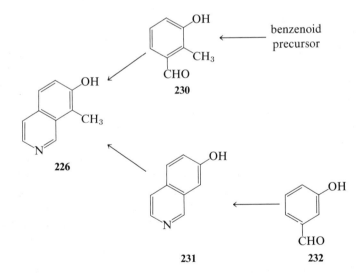

benzenoid
precursor

230

226

231   **232**

ring is fused in the *meta* and *para* positions to the hydroxyl group. The electronic features of aromatic substitution suggest that the carbon-to-carbon bond *para* to the hydroxyl group will be the more readily formed of these two bonds. Therefore, the most efficient way to attach the nitrogen-containing ring to the hydroxybenzene ring is through cyclization of an appropriate side chain into the *para* position. This implies that we begin the synthesis with a phenol that bears the rudiments of the side chain in the *meta* position. An appropriate such compound is a *meta* hydroxybenzaldehyde derivative.

We must now make a decision as to when the methyl group should be attached. Is it better to prepare 3-hydroxy-2-methylbenzaldehyde (**230**) from benzenoid precursors and then add the nitrogen-containing ring, or can the methyl group be added after construction of the *iso*quinoline nucleus? The difficulties which attend the synthesis of 1,2,3-trisubstituted benzene compounds and the availability of an elegant methylation reaction for β-naphthol-type molecules suggest that the latter approach is better. The key, of course, is the methylation reaction, for both aldehydes **230** and **232** are suitable for *iso*quinoline construction.

Inasmuch as phenols are the enolic forms of cyclic dienones, we can expect them to undergo carbonyl reactions which require enolate intermediates. Indeed, the aldol-type condensation of phenol and formaldehyde leads to a commercially important polymer (Bakelite®). Robinson and

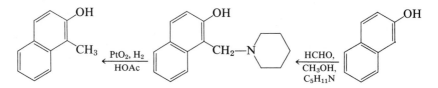

co-workers found that β-naphthol, which reacts selectively in the α-position, will undergo a Mannich reaction with formaldehyde and piperidine very efficiently. α-Methyl-β-naphthol is then readily obtained from this aminophenol by catalytic hydrogenolysis of the benzylic-type nitrogen function. The ease of this substitution sequence offers an attractive method for the introduction of a methyl group in the hydroxy*iso*quinoline **231** formed from *meta*-hydroxybenzaldehyde (**232**), and it precludes consideration of the alternative route.

Application of this Mannich base procedure to the hydroxy*iso*quinoline **231** resulted in the efficient production of the amino derivative **233,** but catalytic hydrogenolysis failed. This behavior is probably a function of the presence in the same molecule of a basic nitrogen function and a phenolic group, for compounds which lack one of these groupings behave normally. The obstacle was overcome by reductive elimination of the piperidine moiety with methoxide-methanol at 220°.

Finally, the synthesis of the hydroxy*iso*quinoline **231** was realized from *meta*-hydroxybenzaldehyde (**232**) and aminoacetal. As noted before, this *iso*quinoline synthesis takes advantage of the greater susceptibility of the position *para* to the hydroxyl group to substitution, and ready cyclodehydration is observed between the aromatic ring and the side chain aldehyde. Thus we have provided the final link to commercially available materials,

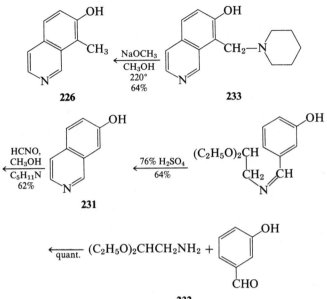

and we have rationalized the synthetic scheme with which Woodward and Doering successfully realized the total synthesis of quinine (**201**).

It is fitting to conclude these comments with a line from another work by Nobel Laureate R. B. Woodward[10] that best describes this kind of effort: "There is excitement, adventure and challenge, and there can be great art in organic synthesis."

[10] R. B. Woodward, "Synthesis" in *Perspectives in Organic Chemistry* (New York, N.Y.: Interscience Publishers, Inc., 1956), p. 158.

7

Problems

The following synthetic problems are presented for practice. The list is not exhaustive nor necessarily representative of all types of synthetic situations that will be encountered. Collateral texts should be consulted for additional examples.

There are usually several ways to prepare a given compound, and one of them is not necessarily the "correct" scheme. However, for the sake of providing a scale of reference, pertinent references to the chemical literature wherein a suitable synthesis may be found are included after each substance. In cases where no reference is given, the substance is either readily available, and hence an exercise, or the synthesis has not yet been recorded.

1. Prepare the following materials from methanol, ethanol, acetylene, and suitable inorganic reagents.

(a) $\underset{\overset{|}{\text{CO}_2\text{H}}}{\text{HO}_2\text{CCH}_2\text{CHCH}_2\text{CO}_2\text{H}}$ [OS, **1**, 272]

(b) $\underset{\overset{|}{\text{OH}}}{\text{CH}_3\overset{\overset{\text{H}_3\text{C}\ \ \text{CH}_3}{|\ \ \ |}}{\text{CHCCH}_3}}$ [Huston and Guile, *J. Am. Chem. Soc.,* **61,** 70 (1939)]

(c) $\underset{\overset{|}{\text{OH}}}{(\text{CH}_3)_2\text{CH}\overset{\overset{\text{CH}_3}{|}}{\text{CC}}(\text{CH}_3)_3}$ [Nasarow, *Ber.,* **69B,** 23 (1936); **70B,** 599 (1937)]

(d) $(\text{CH}_3)_2$ ⌐⌐ ⌐N⌐O [OS, **IV,** 652]
     N
     H

(e) $\overset{\text{CH}_2}{\square}$ with CO, O, CO   [OS, **43,** 27]

(f) $\underset{}{(\text{CH}_3)_2\overset{\overset{\text{CH}_3}{|}}{\text{CHCHCH}_2\text{CHO}}}$ [Barnes and Budde, *J. Am. Chem. Soc.,* **68,** 2339 (1946)]

140

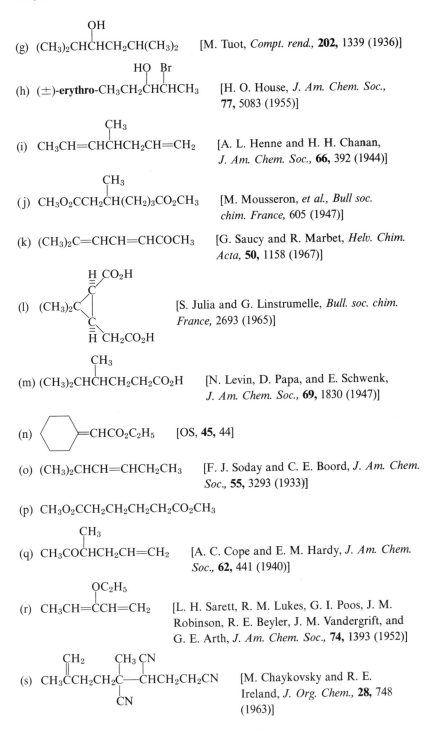

(g)   (CH₃)₂CHĊHCH₂CH(CH₃)₂   [M. Tuot, *Compt. rend.,* **202,** 1339 (1936)]

        $\overset{\text{HO}}{|}\ \overset{\text{Br}}{|}$

(h)   (±)-**erythro**-CH₃CH₂ĊHĊHCH₃   [H. O. House, *J. Am. Chem. Soc.,* **77,** 5083 (1955)]

        $\overset{\text{CH}_3}{|}$

(i)   CH₃CH=CHĊHCH₂CH=CH₂   [A. L. Henne and H. H. Chanan, *J. Am. Chem. Soc.,* **66,** 392 (1944)]

        $\overset{\text{CH}_3}{|}$

(j)   CH₃O₂CCH₂ĊH(CH₂)₃CO₂CH₃   [M. Mousseron, *et al., Bull soc. chim. France,* 605 (1947)]

(k)   (CH₃)₂C=CHCH=CHCOCH₃   [G. Saucy and R. Marbet, *Helv. Chim. Acta,* **50,** 1158 (1967)]

(l)   [S. Julia and G. Linstrumelle, *Bull. soc. chim. France,* 2693 (1965)]

        $\overset{\text{CH}_3}{|}$

(m)   (CH₃)₂CHĊHCH₂CH₂CO₂H   [N. Levin, D. Papa, and E. Schwenk, *J. Am. Chem. Soc.,* **69,** 1830 (1947)]

(n)   =CHCO₂C₂H₅   [OS, **45,** 44]

(o)   (CH₃)₂CHCH=CHCH₂CH₃   [F. J. Soday and C. E. Boord, *J. Am. Chem. Soc.,* **55,** 3293 (1933)]

(p)   CH₃O₂CCH₂CH₂CH₂CH₂CO₂CH₃

        $\overset{\text{CH}_3}{|}$

(q)   CH₃COĊHCH₂CH=CH₂   [A. C. Cope and E. M. Hardy, *J. Am. Chem. Soc.,* **62,** 441 (1940)]

        $\overset{\text{OC}_2\text{H}_5}{|}$

(r)   CH₃CH=ĊCH=CH₂   [L. H. Sarett, R. M. Lukes, G. I. Poos, J. M. Robinson, R. E. Beyler, J. M. Vandergrift, and G. E. Arth, *J. Am. Chem. Soc.,* **74,** 1393 (1952)]

(s)   [M. Chaykovsky and R. E. Ireland, *J. Org. Chem.,* **28,** 748 (1963)]

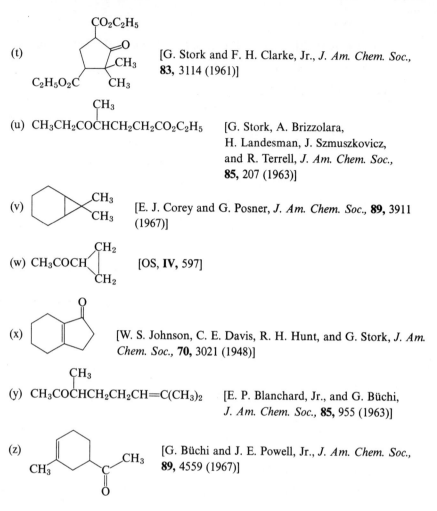

(t) [G. Stork and F. H. Clarke, Jr., *J. Am. Chem. Soc.,* **83,** 3114 (1961)]

(u) $CH_3CH_2COCHCH_2CH_2CO_2C_2H_5$
(with CH_3 on the CH)
[G. Stork, A. Brizzolara, H. Landesman, J. Szmuszkovicz, and R. Terrell, *J. Am. Chem. Soc.,* **85,** 207 (1963)]

(v) [E. J. Corey and G. Posner, *J. Am. Chem. Soc.,* **89,** 3911 (1967)]

(w) CH_3COCH [OS, **IV,** 597]

(x) [W. S. Johnson, C. E. Davis, R. H. Hunt, and G. Stork, *J. Am. Chem. Soc.,* **70,** 3021 (1948)]

(y) $CH_3COCHCH_2CH_2CH=C(CH_3)_2$
(with CH_3 on the CH)
[E. P. Blanchard, Jr., and G. Büchi, *J. Am. Chem. Soc.,* **85,** 955 (1963)]

(z) [G. Büchi and J. E. Powell, Jr., *J. Am. Chem. Soc.,* **89,** 4559 (1967)]

2. Starting with readily available materials, elaborate syntheses for the following compounds.

(a) 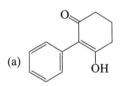 [H. Born, R. Pappo, and J. Szmuszkovicz, *J. Am. Chem. Soc.,* 1779 (1953)]

(b) 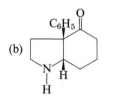 [H. W. Whitlock, Jr., and G. L. Smith, *J. Am. Chem. Soc.,* **89,** 3600 (1967)]

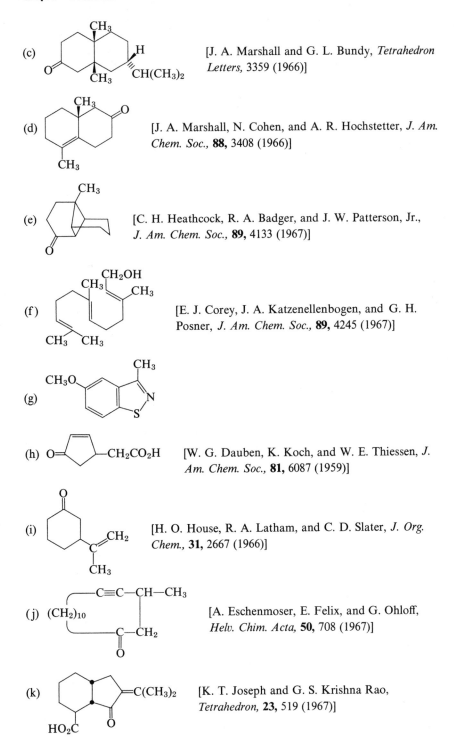

(c)   [J. A. Marshall and G. L. Bundy, *Tetrahedron Letters,* 3359 (1966)]

(d)   [J. A. Marshall, N. Cohen, and A. R. Hochstetter, *J. Am. Chem. Soc.,* **88,** 3408 (1966)]

(e)   [C. H. Heathcock, R. A. Badger, and J. W. Patterson, Jr., *J. Am. Chem. Soc.,* **89,** 4133 (1967)]

(f)   [E. J. Corey, J. A. Katzenellenbogen, and G. H. Posner, *J. Am. Chem. Soc.,* **89,** 4245 (1967)]

(g)

(h)   [W. G. Dauben, K. Koch, and W. E. Thiessen, *J. Am. Chem. Soc.,* **81,** 6087 (1959)]

(i)   [H. O. House, R. A. Latham, and C. D. Slater, *J. Org. Chem.,* **31,** 2667 (1966)]

(j)   [A. Eschenmoser, E. Felix, and G. Ohloff, *Helv. Chim. Acta,* **50,** 708 (1967)]

(k)   [K. T. Joseph and G. S. Krishna Rao, *Tetrahedron,* **23,** 519 (1967)]

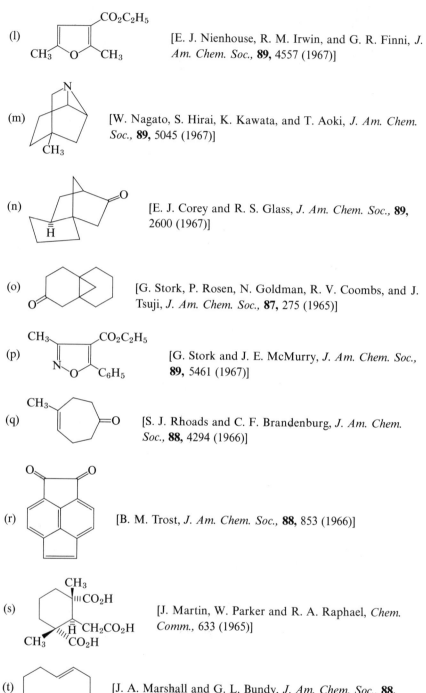

(l)   [E. J. Nienhouse, R. M. Irwin, and G. R. Finni, *J. Am. Chem. Soc.*, **89**, 4557 (1967)]

(m)   [W. Nagato, S. Hirai, K. Kawata, and T. Aoki, *J. Am. Chem. Soc.*, **89**, 5045 (1967)]

(n)   [E. J. Corey and R. S. Glass, *J. Am. Chem. Soc.*, **89**, 2600 (1967)]

(o)   [G. Stork, P. Rosen, N. Goldman, R. V. Coombs, and J. Tsuji, *J. Am. Chem. Soc.*, **87**, 275 (1965)]

(p)   [G. Stork and J. E. McMurry, *J. Am. Chem. Soc.*, **89**, 5461 (1967)]

(q)   [S. J. Rhoads and C. F. Brandenburg, *J. Am. Chem. Soc.*, **88**, 4294 (1966)]

(r)   [B. M. Trost, *J. Am. Chem. Soc.*, **88**, 853 (1966)]

(s)   [J. Martin, W. Parker and R. A. Raphael, *Chem. Comm.*, 633 (1965)]

(t)   [J. A. Marshall and G. L. Bundy, *J. Am. Chem. Soc.*, **88**, 4291 (1966)] (u)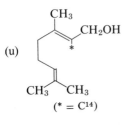

[P. Loew, H. Goeggel, and D. Arigoni, *Chem. Comm.,* 347 (1966)]

(* = C¹⁴)

(v)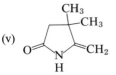

[E. Bertele, H. Boos, J. D. Dunitz, F. Elsinger, A. Eschenmoser, I. Felner, H. P. Gribi, H. Gschwend, E. F. Meyer, M. Pesaro, and R. Scheffold, *Angew. Chem.,* **76,** 393 (1964)]

(w)

[unknown, but see J. A. Marshall and H. Faubl, *J. Am. Chem. Soc.,* **89,** 5965 (1967), and J. R. Wiseman, *ibid.,* **89,** 5966 (1967)]

(x)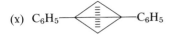

(y)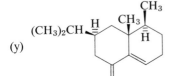

(z)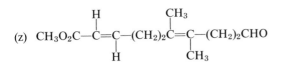

3. Devise a documented synthetic scheme for the following compounds. Cite a literature analogy for each of your proposed steps, and give the specific reaction conditions and yield observed. Give the source of your starting material together with the supplier and price.

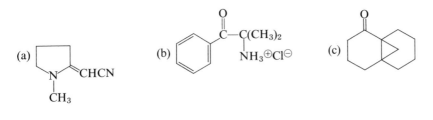

(d)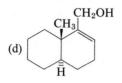

(e)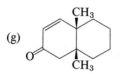

(f)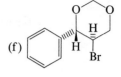

(g)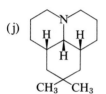

(h)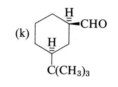

(i) $CH_3O_2C(CH_2)_3\overset{\underset{|}{D}}{C}=\overset{\underset{|}{D}}{C}(CH_2)_3CO_2CH_3$

(j)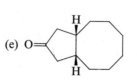

(k)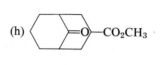

(l)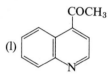

Index